Basic HOUSE WIRING

Basic HOUSE WIRING

MONTE BURCH

Popular Science

Sterling Publishing Co., Inc.   New York

Library of Congress Cataloging-in-Publication Data

Burch, Monte.
  Basic house wiring.

  (Popular science)
  Includes index.
    1. Electric wiring, Interior—Amateurs' manuals.
  I. Title. II. Series: Popular science (Sterling
Publishing Company)
  TK9901.B87   1987   621.319′24   87-10205
  ISBN 0-8069-6516-9 (pbk.)

     7 9 10 8 6

First published in paperback in 1987 by Sterling Publishing Co., Inc.
387 Park Avenue South, New York, N.Y. 10016
Originally published in hardcover by Grolier Book Clubs, Inc.
Copyright © 1982 and 1975 by Monte Burch
Distributed in Canada by Sterling Publishing
℅ Canadian Manda Group, P.O. Box 920, Station U
Toronto, Ontario, Canada M8Z 5P9
Distributed in Great Britain and Europe by Cassell PLC
Artillery House, Artillery Row, London SW1P 1RT, England
Distributed in Australia by Capricorn Ltd.
P.O. Box 665, Lane Cove, NSW 2066
Manufactured in the United States of America
All rights reserved

Reprints and excerpts used with permission from the 1981
National Electrical Code, © 1980, National Fire Protection
Association, Boston, MA 02210.

To

R.J. De CRISTOFORO,

JAY HEDDEN,

and

JOHN W. SILL

the people
who got me started

Contents

Preface

Today more and more homeowners are becoming do-it-yourselfers. Many men and women who thought they would never pick up a hammer are becoming adept at fixing screen doors, working on leaky faucets, and installing or repairing electrical wiring. Some of these people are doing these chores merely for fun, but the majority are learning how to work on their own homes out of necessity.

This book is written primarily for the beginner who is planning to build a new home, or add on rooms, or update the wiring in an older home. Step-by-step explanations show exactly how to go about these jobs.

The information in this book is based on the 1981 *National Electrical Code* book, and naturally it doesn't include all the Code's electrical engineering data. *This book is not meant to rival or supersede any code book, either national or local,* but instead is intended to be used in conjunction with the codes to help the homeowner understand and make electrical installations.

Many people helped me put this book together. First, I would like to thank my wife, Joan, who does all the final manuscript typing, checking and rechecking of my facts—and in general just keeps me rolling. Special thanks go to Jay Hedden of *Workbench* magazine and Larry Woods of *How-To* magazine for allowing me to reuse some photographs from my previously published articles. A very special thanks to James Hollenbeck, an electrician friend who let me follow him from job to job to shoot many of the photos for the book.

I am indebted to the National Fire Protection Association (NFPA) for providing technical review of the text and many illustrations and for granting permission to excerpt from their Code.

I would also like to thank the following fine companies for supplying photographs, drawings, charts, materials, and information: Thomas Industries, Inc.; NuTone Division, Scovill Mfg. Co.; Bell Electric Co.; Daniel WoodHead Co.; Ideal Industries,

Inc.; Warren & Goodin Architects-Engineers; Leviton Industries; Cable Electric Products, Inc.; Bryant Electric, Division of Westinghouse Electric Corp.; Bussman Manufacturing, a McGraw-Edison Company Division; Raco, All-Steel Equipment Inc.; Little Giant Pump Co.; Allen-Stevens Conduit Fittings Corp.; General Electric Co.; Intermatic, Inc.; Purex Corporation, Pool Products Division; Illuminating Engineering Society, publisher of *Lighting Design & Application*.

Monte Burch
Humansville, Missouri

Basic HOUSE WIRING

1

Electricity for the Home

Wiring your own home is actually fairly simple, but to be able to do a proper job, you should understand how electricity is produced, how it works, how it is delivered to your home, and how it operates in your home.

TRANSMISSION

The electricity that enters your home is called "AC" or alternating current, which is different from "DC" or the direct-current electricity used in your automobile. Alternating current flows in one direction, then reverses, and flows back at sixty cycles per second. Alternating current is used instead of direct current because it is more easily controlled when transmitted over long distances.

After the electricity is generated, either by hydroelectric dams, coal furnaces, or nuclear reactors, it is transmitted across country by extremely high-voltage power lines to various regional or town substations. When electricity is transmitted through a line, a gradual drop in the voltage results. This happens even on the short-distance runs of

the circuits in your home, although this voltage drop is extremely slight.

Where electricity is to be transmitted for long distances, a "step-up" transformer steps up the voltage and proportionately reduces the amperage. At each destination, a "step-down" transformer reduces the high voltage electricity to the 120 or 240 volts required for normal household operation. If you follow the electrical service lines from your

The nation's electric power is generated by hydro-electric dams, as well as by coal furnaces and nuclear reactors.

Electricity from generators is transmitted through high-voltage lines to substations like this one.

Step-down transformers on utility poles reduce voltage to desired levels for use in individual homes.

house, you will find a step-down transformer located on the utility company pole which supplies your house.

ELECTRICITY

Electric current flowing through a wire is actually a movement of electrons. Although you can't weigh or see the electrons, they can be measured. The plumbing in your home functions much like your electric system. Water flowing through the plumbing is measured in gallons per minute, and the electrical current running through your lines is calculated in "coulombs per second." One coulomb per second equals 1 ampere. And coulombs are more often called "amps." The amount of pressure of the water in your plumbing is measured in pounds per square inch. The amount of "pressure" in electric current is measured in volts. When you multiply the number of amps a device consumes by the voltage in the power line, you come up with watts—the total amount of energy in a circuit at a given moment. (Amps × volts = watts.)

As mentioned before, the high-voltage electricity from the utility company power lines is converted to the normal 120 or 240 volts which is brought to your house by the service wires. These are connected to your house wiring at the service entrance, which consists of a service head and the wires running into the service entrance panel inside the house. Regarding voltage, some people get confused because for years house voltages have been called 110 or 220. Now you hear of 115 and 230, or 120 and 240 volts. This is the voltage supplied by your local power company. Most voltages supplied today are 120 and 240, but check with your local company as to what voltage you are getting. The nominal difference in the voltages between 110–220 and 120–240 does not really affect your wiring much.

SAFETY

The main thing to remember about electricity is that poor or wrong materials can cause fires, property damage, injury or even death. *The first safety precaution you must take before attempting any electrical work is to shut off the current coming into the house.* In electrical theory, the black and red wires coming in from the service entrance are called "hot" wires, and the white wire is called neutral. But the white wire is always on the return side of the circuit, and current flows through the white just the same as it flows through the other two wires. Or you may encounter two black wires instead of a black and a red.

Even if you shut off power at the service entrance panel by turning off the main circuit breaker, or pulling the main fuse, the electricity will still come into the panel on the service entrance side of the fuse or circuit, and those wires will still be hot. So treat them carefully, and don't touch them. *Remember to always connect the black wires to the black wires and the white wires to the white wires. Never break a white wire for a connection, because it must be continuous to provide the "ground."* This rule may be broken only in specified circuits, as explained in Chapter 7.

THE UTILITY COMPANY

The first step in house wiring, whether you plan to install wiring in a new house or to rewire an older one, is to consult the local utility company. They will tell you where to install the meter head and service entrance for the best location of the service drop wires. This keeps the wires from crossing someone else's property and avoids tree limbs or other obstacles. In addition each utility company will handle the installation a bit differently. In most instances the homeowner provides and installs all the materials up from the service panel to the meter head, and then from the meter head up to the service entrance. The utility company then installs the service drop lines to the entrance head wires, and installs the meter. The utility company will also need to know what size service you intend to install—100 amp, 150, and so on. Let them know about your intended wiring job, and they will normally send a man to look over your situation and give you advice for a proper hookup to their service wires.

READING YOUR METER

The meter the utility company installs measures the amount of electricity you use. By reading your own meter, you can determine exactly how much electricity your

This is a typical meter by which the utility company keeps track of electric consumption.

Using the meter-reading technique described in the text, you should come up with a reading of 08458 kilowatt hours here.

house is consuming. Taking a reading is actually quite easy. The electricity consumed in your home is measured in kilowatt-hours, or "KWH." A meter will have five dials on it. The first dial, which is on the right-hand side, shows units of KWH, the second dial 10s of KWH, the third 100s of KWH, the fourth 1000s of KWH, and the fifth 10,000s of KWH. To read your meter, start at the first dial—the right-most dial. If the pointer is between two numbers, read the lower number. Continue reading, recording from right to left.

THE CODES

Whenever you encounter "the Code" in this book, it will refer to the *National Electrical Code,* a book put out every three years by the National Fire Protection Association. The Code is a guide for the proper use of materials and for correct, safe installations during an electrical wiring job. The National Association does not have the authority to enforce the rules in the Code, but most local codes carefully follow it and, in some cases, are even more strict. And local authorities will enforce the rules they have established or accepted.

You should obtain a copy of the over 600-page Code if you plan to do any extensive wiring. You might be able to borrow a copy from your utility company. Or you may be able to borrow one from your local

library. Or you can order the Code from the National Fire Protection Association, Batterymarch Park, Quincy, MA 02269.

There are a lot of misconceptions concerning the use of the Code. Many people feel it is too big and complicated for the average guy to understand. But although the

From the periodically revised *National Electrical Code,* all local codes books are derived. You'll find the Code of great help if you plan any extensive wiring jobs.

Local code books provide special requirements which you'll have to comply with for okays from local electrical inspectors.

Code is large, a great deal of it doesn't concern home wiring. The truth is that it is quite easy to read, if you understand how it is put together. Instead of a subject listing by page number, you get a "Table of Contents" that lists subjects by articles and sections. If you wanted to know the correct number of receptacles to install in a room, you would look in the "Contents" under Chapter 2, "Wiring Design and Protection." There you would find that Article 210, page 36, tells the complete required number of receptacles for each room, as well as where they should be located. Again the Code is not law. Yet it may be interpreted as such by the county or municipality in which you live.

Whether you plan to do minor or major wiring, a copy of the local code is a must.

Most utility companies are glad to furnish the local code free.

PERMIT

In some areas you must pay a permit fee. An electrical inspector will then come around during construction or installation to insure the work is done properly and then again to examine the final job. *In a small percentage of localities, homeowners are not even permitted to do their own work.* On the other hand, some local authorities will give you all the advice they can, perhaps giving you their local code book as well as a small pamphlet which gives hints and tips on proper installations. Policies differ from county to county, or even from town to town. So make sure you understand the local code rules and follow them carefully.

UNDERWRITERS LABORATORIES

The Underwriters Laboratories Inc., or more simply called "UL," is a nonprofit organization devoted to testing and establishing standards for electrical equipment. A common misuse of the UL label is to state that UL has "approved" an item. The UL doesn't approve items, but instead it "lists" them, which means each product listed has passed UL tests and meets minimum safety standards. Look for the UL listed seal or stamp

Look for the UL stamp on all electrical materials. This signifies that the Underwriters Laboratories Inc. has tested the product and lists it as meeting minimum standards for quality and safety.

on all electrical equipment you purchase. Normally an electrical inspector will approve only materials listed by UL. Manufacturers normally show UL listing prominently on their products and packaging.

Incidentally, some items such as door-bell wires are not listed because they normally run on low voltage. Door chime transformers, however, are listed.

GLOSSARY

To help you further understand electricity and house wiring, here is a glossary of the common terms and their meanings:

Alternating Current (AC). A current alternating many times a second. It flows back and forth first in one direction, then in the other.

Ampacity. The current-carrying capacity of the electric conductor expressed in amperes

Ampere. A unit measure of electrical flow based on the number of electrons per second flowing past a given point

Appliance. Equipment such as clothes dryers, washing machines, food mixers

Appliance, Fixed. An appliance which is built-in or otherwise secured at a specific location

Appliance, Portable. An appliance easily moved, such as a hand iron, coffee maker

Appliance, Stationary. An appliance which is not easily moved to permit normal use. Includes refrigerators and free-standing electric ranges

Approved. Acceptable to the authority having jurisdiction

Bonding. The permanent joining of metallic parts to form an electrically conductive path which will assure electrical continuity and safe flow of current

Branch Circuit. A circuit conductor between the outlet and the final overcurrent device protecting the circuit

Branch Circuit, Appliance. Supplies energy to one or more outlets to which appliances are to be connected; such circuits should have no permanently connected lighting fixtures that are not a part of an appliance

Branch Circuit, Individual. A branch circuit supplying current to only one piece of equipment

Candlepower. A measure of the intensity of light produced by one source. One candlepower approximately equals the light produced by an ordinary candle.

Circuit. Two or more wires which conduct electricity from the source to one or more outlets and back again

Circuit Breaker. A safety device that breaks electricity flow when a circuit becomes overloaded or short circuited

Circuit Breaker, Adjustable. A term indicating the circuit breaker can be set to trip at various values of current and/or time in a predetermined range

Circuit Breaker, Instantaneous Trip. A term indicating no delay is purposely introduced in the tripping action of the circuit breaker

Circuit Breaker, Inverse Time. A term indicating there is a purposely introduced delay in the tripping action of a circuit breaker. This delay decreases as the magnitude of the current increases.

Circuit Breaker, Nonadjustable. A term indicating that the circuit breaker does not have any adjustment to alter the value of current at which the breaker will trip or the time required for operation

Circuit Breaker Setting. The value of current and/or time at which an adjustable circuit breaker is set to trip

Color-Coding or Polarizing. A means of identifying wires by color throughout a system to help assure that hot wires will be connected only to hot wires and that neutral wires will run without interruption back to ground terminal

Conductor. A common term for electric wire

Conductor, Bare. A wire having no covering of electrical insulation

Conductor, Covered. A wire encased within a material that is not recognized by the Code as electrical insulation

Conductor, Insulated. A wire encased in a material that is recognized by the Code as electrical insulation

Connector, Pressure (Solderless). A device establishing a connection between one or more conductors and a terminal by means of mechanical pressure without the use of solder

Cooking Unit, Counter-Mounted. A cooking appliance designed to be mounted in or on a counter and consisting of one or more heating elements, internal wiring, and built-in or separately mounted controls

Cycle. The period of complete alternation of alternating current, consisting of one positive (forward flow) and one negative (backward flow) alternation. Ordinary 60-cycle current has 60 cycles or 120 alternations each second.

Direct Current (DC). Flow of electric current continuously in one direction in a closed circuit

Direct Lighting. A lighting system which delivers the majority of light in one direction without being deflected from the ceiling or walls. This would be any lamp or fixture equipped with glass or metal reflector arranged to reflect light toward the object to be illuminated.

Disconnecting Means. Device and/or devices by which conductors of a circuit can be disconnected from the source of supply

Equipment. A general term that includes material, fittings, devices, appliances, fixtures and apparatus used in electrical installations

Explosion-Proof Apparatus. An apparatus enclosed in a case that is capable of withstanding an explosion of gas or vapor which may occur within it. The case can also prevent ignition of gas or vapor outside as a result of sparks, flashes, or explosion of the gas or vapor within. And the apparatus operates at an external temperature that will not ignite a surrounding flammable atmosphere.

Exposed. Capable of being inadvertently touched by a person. This applies to parts not suitably guarded, isolated, or insulated.

Feeders. All circuit conductors between service equipment, or the generator switchboard of an isolated plant, and the final branch circuit overcurrent device

Filament. A slender thread of material such as carbon or tungsten which emits light when raised to high temperature by electric current (as in an incandescent light bulb)

Fitting. An accessory such as a locknut, bushing, or other part of a wiring system that is intended primarily to perform a mechanical rather than an electrical function

Footcandle. The amount of illumination on a square foot of surface that is everywhere one foot from one lumen of light (as from a candle)

Frequency. The number of complete cycles per second in alternating current circuit. The frequency of a 60-cycle circuit is 60.

Fuse. A safety device that breaks the flow of electricity when a circuit becomes overloaded or short circuited

Ground. A conducting connection between

an electrical circuit or some equipment and the earth. Or it can be a connection to some conducting body that serves in place of the earth.

Grounding Conductor. A conductor used to connect equipment or a grounded circuit of a wiring system to a grounding electrode or to electrodes

Horsepower (HP). A unit of power equal to 746 watts

"Hot" Wires. Power-carrying wires (usually black or red)

Indirect Lighting. A system of lighting that directs all light to the ceiling or to walls, which in turn reflect light to the objects to be illuminated

Insulation. Protective sheathing used over wires to prevent escape of electricity

Kilowatt. 1,000 watts

Kilowatt Hour (KWH). 1,000 watts used for 1 hour

Labeled. An item that is listed and has been tested by an organization such as Underwriters' Laboratories

Lighting Outlet. An outlet intended for the direct connection of a lampholder, a lighting fixture, or a pendant cord terminating in a lampholder

NEC. The *National Electrical Code,* the electrician's bible, revised periodically by the National Fire Protection Association

OHM. A unit of electrical resistance

Outlet. A device permitting the tapping of electricity for lights or appliances

Oven, Wall-Mounted. A cooking oven designed to be mounted in or on the wall and consisting of internal wiring, built-in or separately mounted controls, and one or more heating elements

Overcurrent Device. A device, such as a fuse or circuit breaker, which limits the amps in a wire to a predetermined amount

Rainproof. An apparatus constructed, pro-tected, or treated to prevent rain from interfering with successful operation

Receptacle. A contact device installed at the outlet into which electric cords can be plugged. A single receptacle is a single contact device with no other contact devices on the same yoke. A multiple receptacle is a single device containing two or more receptacles.

Receptacle Outlet. An outlet where receptacles are installed

Service. Conductors and equipment delivering energy from the electricity supply system to the wiring system of the premises

Service Cable. Service conductors made up to form a cable

Service Conductors. Supply conductors that extend from a street main or from transformers that service household equipment

Service Drop. Overhead service conductors from the last utility pole or aerial support to and including the splices connecting to service entrance conductors at the building

Service Entrance. A combination of intake conductors and equipment, including service entrance wires, an electric meter, a main switch or circuit breaker, and the main distribution or service panel

Service Entrance Conductors, Overhead System. Service wires between terminals of service equipment and a point usually outside the building—there joined by tap or splice to the service drop

Service Equipment. Necessary equipment, usually a circuit breaker or switch, fuses and accessories located near the point of entrance of supply wires to a building. This is intended to constitute the main control and means of cutting off the supply.

Service Lateral. Underground service wires

between the street main and the first point of connection to the service entrance wires in a terminal box or meter inside or outside the building wall. When there is no terminal box or meter, the point of connection shall be the point of entrance of the service wires into the building.

Service Panel. The main panel or fuse cabinet through which electricity is brought into the building and distributed to branch circuits. This contains the main disconnect switch and fuses or circuit breakers.

Short Circuit. An improper connection between hot wires or a hot and a neutral wire

Switch. A device that can break the flow of current

Switch, 3-Way. Switches used in pairs that control the same light from two different points. For detailed descriptions, see Chapter 7.

Underwriters Laboratories Inc. A nationally recognized organization that tests all types of wiring materials to make certain they meet minimum standards for safety and quality

Volt. A unit measure of electrical pressure

Voltage Drop. A term indicating the voltage loss which occurs when wires are overloaded

Watt. A unit showing current drain which accounts for both voltage and amperage. Watts equal amps multiplied by volts. Examples: One ampere at pressure of 1 volt equals 1 watt. And 6 amps $\times$ 120 volts = 720 watts.

Weatherproof. Constructed or protected in such a manner that exposure to weather will not interfere with operation

2

Determining Your Needs

Before you can start the mechanical wiring of your home, you must first figure how much electrical power you are now consuming, and more important, how much you will be consuming. This insures that your home will be adequately wired. This also provides you with a more convenient home and a safer one as well. For inadequate wiring is one of the major causes of electrical fires in homes.

Even though a house may be "adequately" wired according to some local code, it still may be inefficiently wired. The idea is to plan your wiring for the most efficient use of electricity as well as for an adequately wired system. In an efficiently wired house there should be little need for extension cords. Also, a person should be able to walk into any room throughout the house, down into the basement and back up, without having to retrace his steps to turn off lights. This is made possible by the correct number of lights, the proper types of lights, and the use of 3-way or even 4-way switches. (Three-way switches are used in pairs to control light from two different points; for

three or more points of control, 4-way switches are used in conjunction with the pair of 3-way switches.)

A house that is adequately wired permits plugging in an appliance without having to unplug another one. In other words, a house that has the proper service, the correct number of circuits, and plenty of lights and receptacles is much more pleasant to live in.

The consumption of electricity in the home has more than tripled in the past 20 years, yet very few homes are wired to cope with this extra burden. Regardless of whether you're wiring a new home or rewiring an older home, make sure the wiring is adequate to handle not only today's electrical needs, but also the needs for the next 15 to 20 years. There are several factors to consider in making sure you have an adequately wired and efficiently operating electrical system.

SERVICE ENTRANCE

The service entrance to your home must have a large enough capacity to serve your needs. In the past years a No. 6, 3-wire 60-amp service entrance was considered

Here a faulty connection in a ceiling junction box caused a fire that was doused quickly.

An old-style light fixture, wired without any type of insulation, resulted in this devastation.

adequate. And in the case of a small house, it might still be. But today most local codes require a 100-amp service-entrance as the minimum. It utilizes a No. 2 or No. 3, three-wire installation with RHW or THW insulation, which can be used in wet or dry locations. RHW consists of wire, with a rubber insulation sheathed in a moisture- and fire-resistant braided cover. Type THW consists of flame-retardant, moisture- and heat-resistant thermoplastic insulation. This type of service entrance will be adequate for most homes with up to 3,000 square feet of floor space.

If you are using "full-electric" appliances, and have a shop in your basement, or electric house-heating, a 150- or 200-amp service would probably be your best choice. Wire size 1/0 or 2/0 copper or No. 2/0 or 4/0 aluminum wire with Type RHW insulation in a 3-wire cable assembly is nor-

This dangerous but common method of overloading a circuit indicates an inadequate number of outlets.

For most small homes today, a 100-amp service panel is the minimum allowed by code.

mally used for a 150- or 200-amp service, respectively. Granted, this larger service costs quite a bit more because of the size of the service panel and the size of the wires, but the cost in these materials is more than made up through the more efficient use of electricity. With this type of service entrance, you can run more circuits with fewer units on each circuit and less voltage loss on each circuit.

Just to get some idea of how much electricity is needed in modern homes, con-sider the following chart showing the various heavy-duty appliances and the watts they consume:

Dishwasher	1800 watts
Fuel-fired furnace	800
Garbage disposal	900
Central airconditioning	5000
Automatic washer	700
Built-in room heater	1600
Freezer	350
Water pump	300 to 700
Water heater	2500 to 4500
Automatic dryer	4500
High-speed dryer	8700

When you start counting all the other small appliances such as television, radio, toaster, roaster, plus the light circuits and receptacle circuits, the amount of electricity consumed in the average modern home is high. The following chart gives the watts consumed by most smaller appliances.

Appliances	Watts	
Incandescent lamp	10 and up	
Fluorescent lamp	15 to	60
Christmas tree lights	20 to	150
Clock	2 to	3
Radio	40 to	150
Television	200 to	400
Ultraviolet sun lamp	275 to	400
Infrared heat lamp		250
Heat pad	50 to	75
Electric blanket	150 to	200
Electric razor	8 to	12
Slide or movie projector	300 to	500
Portable heater	1000 to	1500
Portable fan	50 to	200
Room airconditioner	800 to	1500
Sewing machine	60 to	90
Vacuum cleaner	250 to	800
Refrigerator	150 to	300
Steam or dry iron	660 to	1200
Hot plate (each burner)	600 to	1000

Range (oven and all burners "on")	8000	to	14000
Range top	4000	to	8000
Wall-mounted oven	4000	to	5000
Toaster	500	to	1200
Coffee maker	500	to	1000
Waffle iron	600	to	1000
Roaster	1200	to	1650
Rotisserie	1200	to	1650
Deep-fat fryer	1200	to	1650
Frying pan	1000	to	1200
Blender	500	to	1000
Electric knife			100
Mixer	120	to	250

The drawing on page 14 depicts an ideal wiring scheme for an entire house and shows the various appliances and lights. You will note a 60-amp feed panel or subpanel to the home workshop and an outbuilding. Here the home workshop circuit is included in the subpanel. With this, the heavy motors won't overload the rest of the house.

SUFFICIENT WIRES THROUGHOUT THE HOUSE

This doesn't refer to the number of circuits—rather to proper size wires. You'll find a discussion of the wires for the service entrance in Chapter 8. Today's modern wires are either size No. 12 or the smaller No. 14, depending on local codes and inspectors. The wires to be used on the individual heavy-duty appliance circuits are discussed in Chapter 11.

ADEQUATE NUMBER OF CIRCUITS

Regarding circuits, it's better to have too many than not enough. In fact, you almost can't have too many circuits. One important thing is that lights in the house should not all be on the same circuit. Thus, if you have trouble with one circuit, blowing a fuse or tripping a circuit breaker, at least some of the lights in the house will be operable.

In the illustration on page 14, the general purpose circuits 1 through 5 are 120-volts, 20-amp circuits (they could have been 15-amp circuits and would have complied with the Code) and supply 500 square feet of floor space each. They supply power for all the lighting outlets and the receptacle outlets except for those receptacle outlets in the kitchen and laundry. These areas should be supplied by separate circuits.

Circuits 6 and 7 are appliance circuits for the laundry and the kitchen. Circuit 6 is a 20-amp, 120-volt circuit to the laundry outlets. Note: This circuit is not for an automatic washer and dryer, but is for smaller appliances such as irons or clothes washers. Kitchen circuits should have plenty of receptacle outlets. At least one outlet should be installed at each counter work space wider than 12 inches. Kitchen receptacles must, above all, be grounded *duplex* receptacles.

Circuits 9 through 23 are separate appliance circuits for the various heavy-duty appliances. Circuit 9 is a 120-volt circuit to a fuel-fired furnace. Circuit number 10-12 supplies power to a central airconditioner and comprises a 240-volt circuit. Circuit 8 is a 120-volt circuit to a garage or small workshop. Note: For larger home workshops you will need a larger circuit, or more circuits, including a 240-volt circuit.

Circuit 13–15 is a 50-ampere, 240-volt circuit to the electric range. Number 16–18 is a 30-amp, 240-volt circuit to the water heater. Number 19–21 is a 50-amp, 240 circuit to the automatic washer and dryer. Note: You may need two separate circuits for the appliances. No. 22 is a 20-amp 120-volt circuit for the dishwasher. No. 23 is a 20-amp, 120-volt circuit for the disposal. No. 24 is a 60-amp, 240-volt subpanel

CIRCUIT DIAGRAM

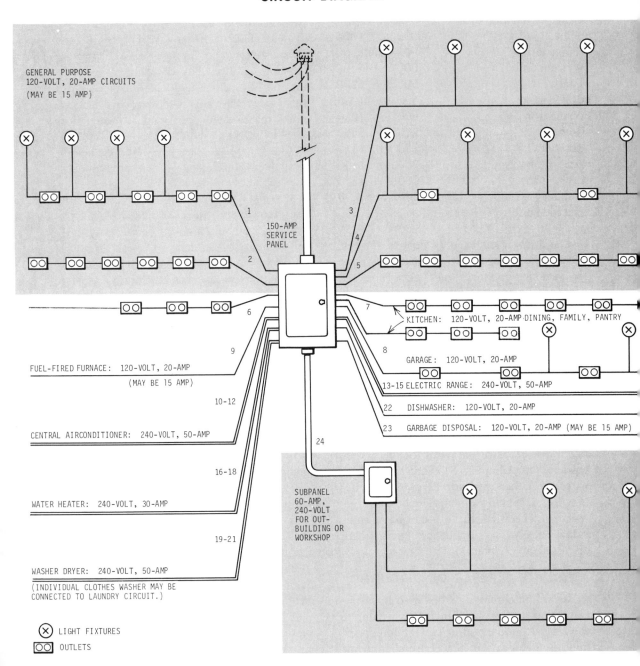

GENERAL PURPOSE
120-VOLT, 20-AMP CIRCUITS
(MAY BE 15 AMP)

1

2

3

4

5

6

7

8

9

150-AMP
SERVICE
PANEL

KITCHEN: 120-VOLT, 20-AMP DINING, FAMILY, PANTRY

GARAGE: 120-VOLT, 20-AMP

13-15 ELECTRIC RANGE: 240-VOLT, 50-AMP

22 DISHWASHER: 120-VOLT, 20-AMP

23 GARBAGE DISPOSAL: 120-VOLT, 20-AMP (MAY BE 15 AMP)

FUEL-FIRED FURNACE: 120-VOLT, 20-AMP
(MAY BE 15 AMP)

10-12

CENTRAL AIRCONDITIONER: 240-VOLT, 50-AMP

16-18

WATER HEATER: 240-VOLT, 30-AMP

19-21

WASHER DRYER: 240-VOLT, 50-AMP
(INDIVIDUAL CLOTHES WASHER MAY BE
CONNECTED TO LAUNDRY CIRCUIT.)

24

SUBPANEL
60-AMP,
240-VOLT
FOR OUT-
BUILDING OR
WORKSHOP

LIGHT FIXTURES

OUTLETS

which could be used for a separate building or for a shop.

In addition to proper size wiring and the proper number of circuits, you'll need the proper number of receptacles, and they should be located in the right places for a convenient system. If receptacles are placed properly, you can do away with the potential danger of extension cords. Wall receptacles should be no more than 12 feet apart, measured horizontally, along walls, dividers, or sliding panels. Floor receptacles can be counted in the 12-foot wall requirement if they are located close to the wall. Generally, wall receptacles should be spaced equally. Each counter space wider than 12 inches in the kitchen and dining room must have a receptacle, and the wall by the bathroom wash basin gets one. You must have at least one outdoor receptacle per family dwelling. The Code also requires that all bathroom and outdoor receptacles must be protected by ground-fault circuit interrupters.

LIGHTING OUTLETS

Normally every room in the house should have a ceiling light, except for the living room or bedrooms, which are usually lighted by lamps. Again, lights are another thing you just can't seem to get enough of. In addition to installing lights in all rooms, make sure there is plenty of lighting in places such as hallways, closets, and the basement stairs.

LIGHTING FOR INDIVIDUAL ROOMS

The living room is normally made up of lamps. Today's most modern lamp fixtures will use 100- to 300-watt bulbs which provide plenty of light if the lamps are properly arranged around the room. In addition, there is normally at least one 3-way lamp which offers a 3-way bulb of different wattages.

In some instances you may wish to place some of the lamps so they can be plugged into receptacles which may be switched on and off, enabling you to turn on the lights

without walking across a dark room. Circuit installation is described in Chapter 7.

Sometimes the dining room is inadequately wired and lit because of the special-occasion, low-light "atmosphere" many people wish to create. The secret is to install a chandelier with a dimmer switch. With this type of installation, plenty of lighting is available when you need it, yet you can dim it to provide "candlelight" too. (For installation instructions, see Chapter 7.) The ceiling light in the dining room may not be in the center of the room and should be located directly over the dining table. Again 3-way switches on either side of the room will help in making the lighting more convenient. In addition to the ceiling chandelier or light, most dining rooms also have wall bracket lights which put out little light, but provide "atmosphere."

A sun room, den or recreation room should be provided with plenty of light. Situate ceiling fixtures so they provide plenty of

In the dining room, a dimmer-controlled chandelier in league with overhead down lights and indirect lighting can produce the desired "atmosphere." Photo by G.E.

light at the most-used portions of the room, as well as provide good overall lighting.

Kitchen lighting should be given extra attention. There should be plenty of light from a good ceiling fixture. For best lighting, consider one of the recessed fluorescent fixtures which gives out an even, soft glow. In addition to the ceiling light, install a light over the sink and an additional light over the stove. If you have a pass-through bar between the kitchen and dining room, you may also wish to install a combination heat-light unit to keep food warm and provide light.

All lights in the kitchen should have switches, and again 3-way switches are a great help. The Code prohibits your pluging kitchen lights into small-appliance receptacle outlets.

You may also wish to install small fluorescent bulbs in or under cabinets. These can be equipped with a touch switch which turns the light on when the door is opened, and off when it is closed.

A kitchen wall clock can be mounted with its cord pushed into an outlet that the clock itself conceals. To conceal the cord, cut it to only a few inches in length.

The bedrooms may have ceiling lights, as well as lamps which may be controlled by wall switches.

If possible, all clothes closets should have lights in them. Some closets, however, may be too small to safely install lighting fixtures in them. There must be at least 18 inches of space between the fixture and any combustible material stored in the closet. Here you can use the simple pull-chain type of fixture. But the metal chain should have an insulating link such as a piece of string on the end to prevent a possible shock. As with the kitchen cabinets, a more modern and convenient manner of lighting closets is to use a light with a switch which turns the light on when

In the kitchen, the vital concerns are illumination of work surfaces and locations of receptacles. Photo by Westinghouse.

the closet door is opened and then off when the door is closed.

Because of presence of water, a bathroom can be a fairly hazardous place. Here all lights should be installed with grounds. There should be a ceiling light for general illumination, plus lights on both sides of the mirror to give ample illumination for shaving or applying makeup. The Code requires that ground-fault circuit interrupters be installed on all 120-volt bathroom receptacle-outlet circuits. In addition, the Code requires one grounded receptacle (again protected by a ground-fault circuit interrupter) beside the wash basin. The main idea is to provide personal protection during the use of any type of portable electric appliance in the bathroom because of shock danger.

One unit that can be a real pleasure in the bathroom is the combination heater and light. It provides both the light needed and that extra bit of heat that is so welcome when you first step out of the bathtub or shower.

All porches should be equipped with a good light that illuminates the doorway as well as any steps leading up to the doorway. If the porch is large, perhaps a summer porch, you'll also wish to install lighting for outdoor summer use.

The installation of lighting in basements varies a great deal according to use. If the basement is used only as a laundry and storage room, naturally the lighting requirements will be less, but be sure to have a light at the head of the stairs. If the basement also serves as a shop or a recreation room, be sure you provide plenty of lights.

Attic, hallways, and stairs are common places where falls and serious injuries occur each year, mostly due to inadequate lighting. Make sure each of these areas in your home is properly lighted. On a long stairway, it's a good idea to light both ends so that someone reaching the bottom can easily see the last step. The main idea is to situate the lights so they illuminate each step, instead of just the general area. Again, these lights should be controlled by 3-way switches at each end of the stairway.

Now that you realize the necessity of proper and adequate wiring, here is how you can determine the needs of your own home. The Code gives two methods of sizing service entrance conductors for residential buildings: the standard method and the alternate method.

The standard method requires that 3 watts per square foot be utilized for the general lighting load, without electric range and major appliances. These are computed separately. The first step is to compute the square footage of your house by measuring the outside dimensions. Each two-wire small appliance load is rated at 1500 watts. After these first loads are calculated, a demand factor may be applied. That is, the first 3,000 watts is rated at 100 percent while the remaining

watts of the general lighting and small appliance loads are rated at 35 percent.

Then all major appliances are listed by their nameplate ratings. The calculated load for the electric range must also be included. Then all figures are totaled to find the total load in watts. Divide this figure by the phase to phase voltage (usually 240 volts) to determine the load in amperes so that you can size the service entrance conductors and panel according to the Code.

To better understand these calculations, let's assume that your single-level home measures 24 feet by 55 feet and requires four small-appliance circuits; an 8000-watt electric-range circuit; a 4500-watt electric water heater circuit; a 4500-watt electric clothes-dryer circuit; and a 5600-watt air-conditioning circuit.

The square footage of the home is $24 \times 55 = 1320$ square feet. So the general lighting load is, 1320 at 3 watts = 3960 watts. Four appliance circuits at 1500 watts totals 6000 watts. The two loads combined gives us 9960 watts. Continue the calculation as follows:

Application of demand factor [Table 220-4 (b) of the Code] 3000 watts at 100 percent	3000 watts
9960 − 3000 watts = 6960 watts at 35 percent	2436
Net computed load without range and major appliances	5436 watts
Electric range	8000
Water heater	4500
Clothes dryer	4500
Air conditioner	5600
Total calculated load	28,036 watts

$$\text{Amperes} = \frac{28,036}{240 \text{ (volts)}} = 116.82$$

Since electric services normally are installed in either 100-, 150-, or 200-amp sizes, you should install a 150-amp service. This will also give you some spare capacity.

In the alternate method of calculating residential service-entrance sizes, the general lighting and small appliance loads are determined in the same manner as in the standard method. However, no diversity is taken until all other loads have been determined. Furthermore, no diversity is allowed on the electric range: the nameplate rating must be figured in totaling the loads. All other major appliances are listed by their nameplate ratings and all the loads are totaled. Here's the load calculation by the alternate method:

General lighting load:	
1320 square feet at 3 watts	3960 watts
Small appliance load 4 circuits at 1500 watts	6000
Electric range (nameplate rating)	12,000
Water heater	4500
Clothes dryer	4500
Air conditioner	5600
Total connected load	36,560 watts

The first 10,000 watts must be rated at 100 percent. The remaining load is calculated on the basis of a 40-percent diversity. Therefore,

First 10,000 watts at 100 percent	10,000 watts
Remaining load at 40 percent (36,560 − 10,000 = 26,560 × 0.40 =)	10,624
Total calculated load	20,624 watts

To find the total load in amperes, divide the total calculated wattage by the phase-to-phase voltage.

$$\frac{20,624}{240} = 859 \text{ amperes}$$

In our second calculation, a 100-ampere service would be sufficient for the load.

The Code requires that there be at least two 20-amp, small-appliance grounded-type circuits for each kitchen and dining room, and at least one for the laundry. The lighting of these rooms may not be included in these circuits. Note: The maximum load-carrying capacities of the various wires are specified by code. Make sure you check the number of watts to be used in a circuit and arrange your circuits so that no circuit will become overloaded.

The following chapter, on planning, will illustrate how you can make up your own working drawing. By including the number of electrical appliances and lights you need, you can come up with a well-planned electrical scheme.

3

Plans and Blueprints

Usually, the best way to plan any involved job is to first "put it down on paper," and this is essential when planning an electrical wiring job. You can save hours of working time, plus many dollars worth of materials, by having a good working drawing or blueprint that shows where each receptacle and light fixture is to be installed, as well as where and how the wires will be run to each of these. For instance, you'll know whether to go through the walls or under the floors with your wires. Most local authorities will require that you make up some sort of wiring diagram to submit to the electrical inspector. Such a drawing helps show that you intend to follow local codes and makes it easier for both you and the inspector to insure that the job will be done properly.

READING AND UNDERSTANDING BLUEPRINTS

Let's start with the blueprints used in the construction of a new house. Regardless of whether you're planning to wire a new house or an addition, or to rewire an older house, electrical wiring blueprints or homemade

wiring diagrams are made up basically the same way. Many people think of house blueprints as mysterious sets of drawings that only an engineer or experienced contractor could understand. And sometimes, if they're poorly done, they can be just that! Actually, an ordinary set of house plans from a good reputable company or architect is fairly easy to understand by the average do-it-yourselfer, provided he does one thing: learn the abbreviations and symbols used for the various materials in the house. In a typical blueprint wiring diagram, the locations of the wiring devices are marked with the electrical symbols and abbreviations established by the American National Standards Association.

It's much easier to use these symbols and abbreviations than to try to pictorially represent each item, or to write in its name. The symbols shown on the following page are the standard electrical symbols found on all electrical wiring blueprints or diagrams.

Learn the symbols or keep the chart handy when studying your house blueprints so you can understand fully what goes

Electrical Reference Symbols

ELECTRICAL ABBREVIATIONS

(Apply only when adjacent to an electrical symbol)

Dust Tight	DT
Explosion Proof	EP
Grounded	G
Rain Tight	RT
Recessed	R
Vapor Tight	VT
Water Tight	WT
Weather Proof	WP

ELECTRICAL SYMBOLS

SWITCH OUTLETS

Single-Pole Switch	S
Double-Pole Switch	S_2
Three-Way Switch	S_3
Four-Way Switch	S_4
Key-Operated Switch	S_K
Switch and Fusestat Holder	S_{FH}
Switch and Pilot Lamp	S_P
Fan Switch	S_F
Switch for Low-Voltage Switching System	S_L
Master Switch for Low-Voltage Switching System	S_{LM}
Switch and Single Receptacle	⊖S
Switch and Duplex Receptacle	⊜S

Door Switch	S_D
Time Switch	S_T
Momentary Contact Switch	S_{MC}
Ceiling Pull Switch	Ⓢ
Multi-Speed Control Switch	Ⓜ

RECEPTACLE OUTLETS

Where weather proof, explosion proof, or other specific types of devices are to be required, use the uppercase subscript letters. For example, weather proof single or duplex receptacles would have the uppercase WP subscript letters noted alongside of the symbol. All outlets should be grounded.

Single Receptacle Outlet	
Duplex Receptacle Outlet	
Triplex Receptacle Outlet	
Quadruplex Receptacle Outlet	
Duplex Receptacle Outlet— Split Wired	
Triplex Receptacle Outlet— Split Wired	
250-Volt Receptable Single Phase Use Subscript Letter to Indicate Function (DW-Dishwasher; RA-Range, CD-Clothes Dryer) or numeral (with explanation in symbol schedule)	
250-Volt Receptacle Three Phase	
Clock Receptacle	Ⓒ
Fan Receptacle	Ⓕ
Floor Single Receptacle Outlet	

Floor Duplex Receptacle Outlet	
Floor Special-Purpose Outlet	*
Floor Telephone Outlet-Private	

CIRCUITING

Wiring Exposed (not in conduit)	—E—
Wiring Concealed in Ceiling or Wall	
Wiring Concealed in Floor	
Wiring Existing**	
Wiring Turned Up	
Wiring Turned Down	
Branch Circuit Home Run to Panel Board	2 1

Number of arrows indicate number of circuits. (A number at each arrow may be used to identify circuit number.)***

---

* Use numeral keyed to explanation in drawing list of symbols to indicate usage.

** Note: Use heavyweight line to identify service feeders. Indicate empty conduit by notation CO (conduit only).

*** Note: Any circuit without further identification indicates two-wire circuit. For a greater number of wires, indicate with cross lines, e.g.:

3 wires;

4 wires, etc.

Neutral wire may be shown longer. Unless indicated otherwise, the wire size of the circuit is the minimum size required by the specification. Identify different functions of wiring system, e.g., signalling system by notation or other means.

Electrical Reference Symbols (continued)

PANELBOARDS, SWITCHBOARDS AND RELATED EQUIPMENT

Flush Mounted Panelboard and Cabinet *

Surface Mounted Panelboard and Cabinet *

Switchboard, Power Control Center, Unit Substations (Should be drawn to scale.) *

Flush-Mounted Terminal Cabinet (In small scale drawings the TC may be indicated alongside the symbol.) *

Surface-Mounted Terminal Cabinet (In small scale drawings the TC may be indicated alongside the symbol.) *

Pull Box (Identify in relation to wiring-system section and size.)

Motor or Other Power Controller (May be a starter or contactor.)*

Externally Operated Disconnection Switch*

Combination Controller and Disconnection Means*

* Identify by Notation or Schedule

POWER EQUIPMENT

Electric Motor (HP as indicated)

Circuit Element, e.g., Circuit Breaker

Circuit Breaker

Fusible Element

Single-Throw Knife Switch

Double-Throw Knife Switch

Ground

Battery

Contactor

Photoelectric Cell

Voltage Cycles, Phase   Ex: 480/60/3

Relay

Equipment Connection (as noted)

LIGHTING

	Ceiling	Wall
Surface or Pendant Incandescent Fixture (PC = pull chain)	TYPE / WATTS	SWITCH / PC CIRCUIT
Surface or Pendant Exit Light		
Blanked Outlet	B	B
Junction Box	J	J
Recessed Incandescent Fixtures	O	

Surface or Pendant Individual Fluorescent Fixture

Surface or Pendant Continuous-Row Fluorescent Fixture (Letter indicating controlling switch)

Fixture No.
Wattage

Symbol not needed at each fixture

*Bare-Lamp Fluorescent Strip

RESIDENTIAL OCCUPANCIES

Signalling-system symbols where a descriptive symbol list is not included on the drawing.

Pushbutton

Buzzer

Bell

Combination Bell-Buzzer

Chime

Annunciator

Electric Door Opener

Television Outlet

The electrical symbols are those suggested for use on drawings as prepared jointly by the American Consulting Engineers Council and the Construction Specifications Institute and as published in CSI Document 16015 (June 1973). Reproduced with permission of the Construction Specifications Institute, Washington, DC 20036.

where. When making up your own blueprint or drawing to submit to an electrical inspector, use the symbols as well. They'll speed up understanding for everyone.

In addition to the symbols shown, letter abbreviations may be placed in or near the symbols. These are some of the most common for house and farm wiring:

DT —Dust-tight
EP —Explosion proof
R —Recessed
G —Grounded
RT —Raintight
WT—Watertight
VT —Vaportight
WP—Weatherproof

MAKING YOUR OWN WIRING PLANS

After studying the details of the house wiring blueprint, it's easy to appreciate how important a good plan or drawing is for the installation of a complete wiring job in either an old or a new house. Naturally you won't be interested in making a drawing as elaborate as a blueprint done by a professional architect, but you should strive to make an "accurately sized" plan of the house to be able to determine exactly how much wiring materials you will need.

Use architectural paper which has small squares lined on it. Allow each line to represent either 1 foot or ½ foot, depending on the size of your house and whether or not you can get the entire floor plan drawn on the paper. The larger size drawing will make it easier to figure the amount of materials needed. Measure the outside of the house; then measure the inside of each room; and

roughly draw these walls in position on the graph paper. Also mark significant features such as doors, windows, fireplaces. Once again symbols don't have to be fancy, just simple indications of what they represent. Mark the intended locations of light fixtures and other hardware. (Chapter 2 explains how to determine these needs.) Mark the position of the various fixtures using the standard symbols and abbreviations.

If you wish to speed up the chore, you can purchase a small plastic template from an art supply store which allows you to trace silhouettes of most of the electrical symbols. With all electrical units marked in place, you can determine the best way to run the wires with the least amount of effort and materials. It's a good idea to make the initial drawing on the squared drawing paper, then use pieces of thin tracing paper

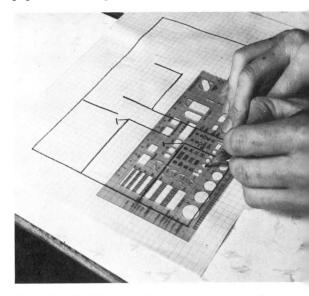

A template like this one helps speed up the task of drawing a wiring diagram.

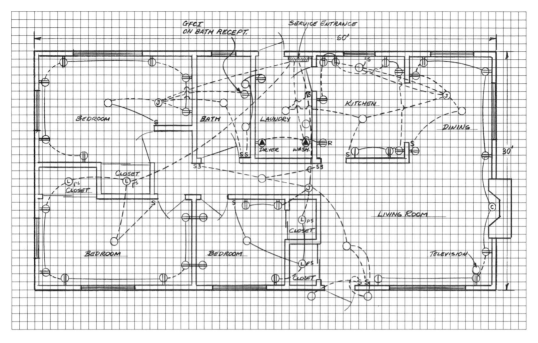

Here's an example of a working drawing you could present to a local inspector.

over this to mark the lines indicating the wires for the various circuits. You'll probably end up changing these circuit lines several times before finally deciding on the best locations anyway. Once you have decided where each item is to be located, make up a final drawing.

Now, by counting the squares, you can determine how many feet of wire will be needed for each run. Remember you must also figure the height of the walls for each run of circuit wire down to a receptacle. Add 18 inches for bends in ceiling and floor plates, as well as about a foot for each connection at an outlet box. By adding all these figures together, you can come up with a fairly accurate estimate of the amount of wire you will need for the entire job.

After placing the items in position on the drawing, according to code with regard to the number of receptacles or lights, you can tally up what you'll need. You can then show this to the electrical inspector, and he will probably suggest some better ways to run wire or to place fixtures.

4

Tools

If you're a do-it-yourselfer or if you have a small workshop in your basement or garage, you probably own most of the tools you'll need for wiring your own home. There are, however, several special tools that make some of the jobs easier and safer as well.

Whenever I have a home remodeling job to do, I always buy the tools I lack and then do the job myself. Usually the cost of the tools is much less than that of hiring someone else to do the job. And when I'm through, I have the tools.

When purchasing tools, buy the best. Quality tools are easier to use, and with proper care, they will last a lifetime. Several of the tools you'll need—such as the conduit bender—might be used for only one job. So rather than buy these one-use tools, you can probably rent them from large hardware stores selling electrical supplies, from mail order houses, or from rental companies.

Knowing how to use each tool properly not only speeds up the job but makes it much safer too. In some cases, professional electricians substitute one tool for many,

thereby saving a great deal of time. But for the do-it-yourselfer, using individual tools designed for specific tasks makes for easier work.

PLIERS

Although an ordinary pair of utility pliers will do the job, it's much easier to use pliers designed for each specific job. Select pliers with good rubber or plastic coatings on them, but don't depend on the coating to protect you from shock. *Always make sure all power is off before working on any electrical circuit or fixture.* Even though the rubber or plastic handles protect your hands from shock, if the day is hot and you're perspiring, water running from your hands down onto the metal portions of the pliers can be fatal. So can a slip of the hand! The coatings on the handles will also make the pliers much easier to use.

Lineman's Pliers

Also called electrician's pliers, these have heavy square jaws which are ideally suited

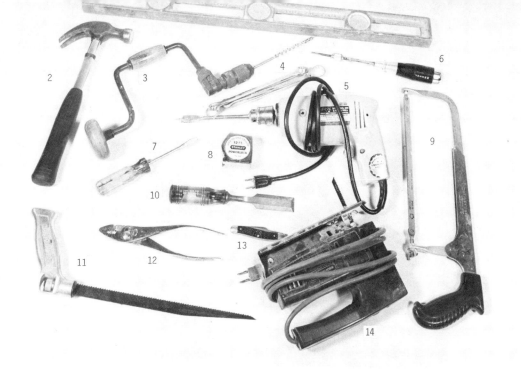

You may already have many of the tools you'll need for electrical work. As you can see from these, carpentry plays an important role. Tools here include 1) level, 2) hammer, 3) brace and bits, 4) wrenches for conduit work, 5) portable electric drill, 6) Yankee drill, 7) screwdriver, 8) tape rule, 9) hacksaw, 10) wood chisels, 11) keyhole saw, 12) utility pliers, 13) pocketknife, 14) saber saw.

Tools you may need to buy include 1) lineman's pliers, 2) toolbag, 3) soldering gun, 4) two-prong tester, 5) combination stripper-crimper, 6) electrician's screwdriver, 7) stud finder, 8) extension bits for brace.

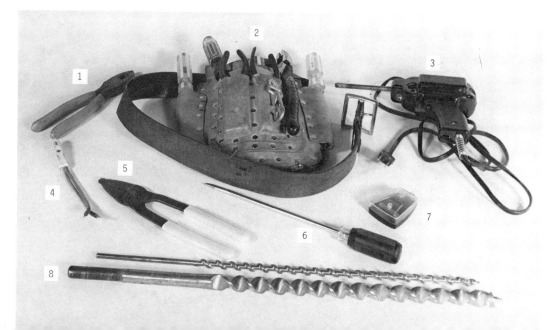

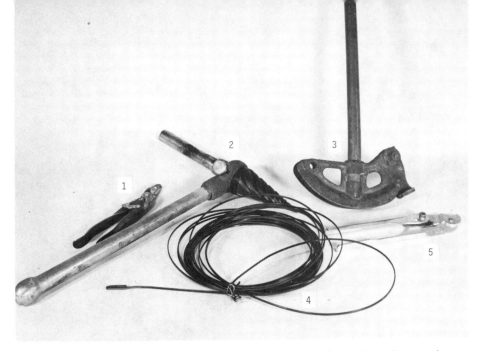

The very specialized and expensive tools can usually be rented from tool rental companies or hardware dealers that supply electrical materials. Shown here are 1) aviation shears, 2) conduit reamer, 3) conduit bender, 4) fish tape, 5) armored-cable tool.

for holding wires while twisting them into splices. They're heavier than utility pliers and provide more leverage and grip. They have a side cutting edge which can be used for cutting heavy wires. Wrapping black plastic electrician's tape around the jaws will help prevent your marring and tearing insulation when making splices.

Side Cutters and End Cutters

Although not necessary for most house wiring, these specialty pliers can come in handy for certain jobs, such as reaching into tight places to snip off pieces of wire.

Needle-Nose Pliers

A good pair of lineman's pliers and a pair of strong needle-nose pliers are actually

Though you can do without them, a pair of lineman's pliers make it easy to cut through heavy wire.

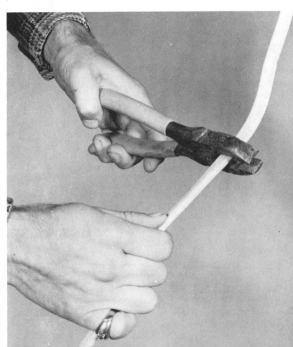

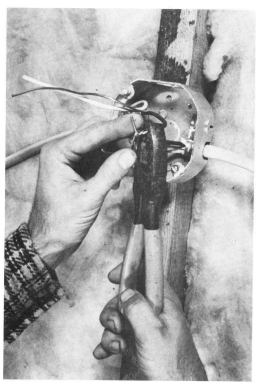
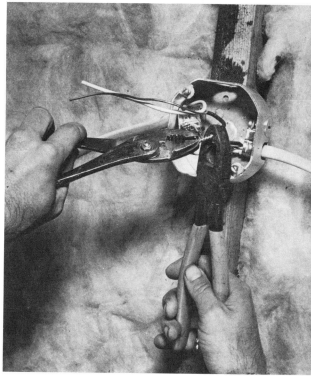

The flat, wide jaws on lineman's pliers clamp wires in place while you start a twist splice, above left. Next, continue the twist with a second pair of pliers, making an even, tight spiral.

Then clip off the jagged edges of the splice using the cutting portion of the jaws.

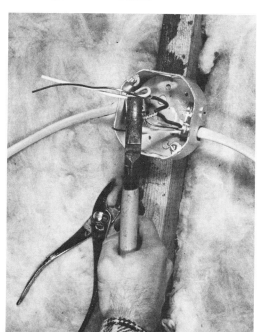

The result is a quality splice that is ready to be soldered or secured with a screw-on connector.

all you will need for practically all electrical work. The needle-nose pliers just can't be beat for making small loops in wire for fitting over such things as receptacle screws. They can also be used to reach into tight places such as boxes or even into a hole cut in a wall to help fish a wire through.

Combination Stripper-Pliers

These are probably the handiest tool you can buy for house wiring and other general electrical work. In addition to their use for gauging and stripping wire, they can be used to crimp on crimp fittings, make small terminal loops, and cut small and medium-size wire. They're the one tool you'll find you just can't get along without. However, their design is such that they don't have the leverage or power for cutting heavy wire, nor can their thin metal ends be used for any heavy leverage, as would be applied when removing knockouts in boxes.

POCKETKNIFE

Just as the old-time carpenter will grudgingly admit that his pocketknife is probably the most versatile tool he owns, most professional electricians use a good sharp pocketknife a great deal as well. A knife with a

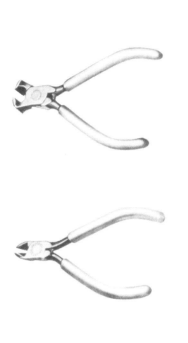

End cutters and side cutters can help you snip wire in difficult-to-reach places.

A pair of needle-nose pliers make it easy to form the terminal loops in wire, and they can reach into places where larger pliers can't.

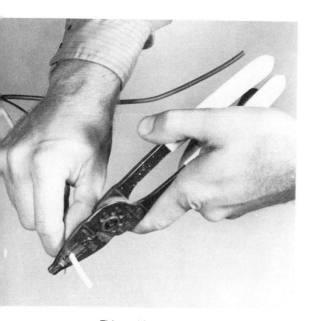
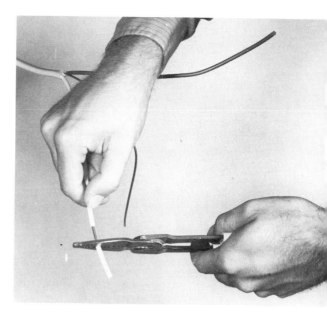

This multi-purpose stripper, crimper, side cutter, pliers is a better investment than single-purpose tools. To strip wire, place it in the proper jaw niche. Clamp the jaws, spin the tool, slip insulation off.

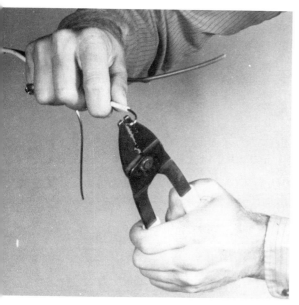
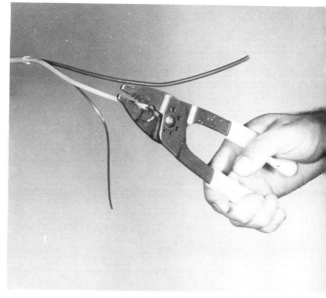

The end of the combination tool can be used for bending terminal loops in the wire.

The center portion of the jaws does the cutting, and the holes behind the pivot are wire gauges.

straight or hook blade, not the traditional curved-up point blade, just can't be beat for slicing away the outer insulation from wire and cutting away the paper insulation strips. Watching a professional make a long slicing cut down the length of a strip of wire, peel it back and cut it off in a second or two, you'll realize how much use a good sharp knife can be. But it must be absolutely sharp or you will pull harder and run the risk of slicing yourself and also of cutting into the inner insulation surrounding the wires. Use a good whetstone and keep the knife as sharp as possible.

STRIPPERS

There are all kinds of strippers on the market, and many electricians like to use the single strippers rather than the combination tool. One type of stripper is much like a small pair of scissors. You merely set the stripper to the gauge wire you wish to strip and clamp it down on the wire. Spin it around the wire and pull off the insulation. One of the more professional types resembles a potato peeler with a guarded hook over the blade. By using a screwdriver you can set the stripper to any gauge wire, push it in place, spin it around the wire and pull off the insulation. This type is excellent for removing a section of insulation from the center of a piece of wire. Make one cut, slide the stripper down the wire, making a slicing cut as you go. Then spin it around the wire to make the end cut, and peel off the insulation. This particular stripper also has a straight-bladed knife on the end for cutting away the outer insulation and paper on larger wires.

SCREWDRIVERS

Almost every household has an assortment of screwdrivers, but few are ideally

A flat blade on a pocketknife cuts plastic-sheathed cable better than a curved blade.

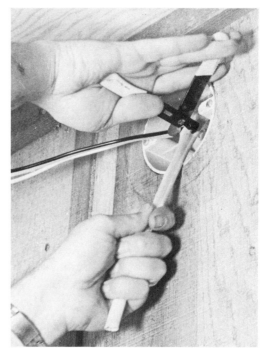

A wire stripper strips any wire and also removes sheathing and insulation from plastic-sheathed cable.

A good plastic-handled screwdriver is a must. Make sure the blade fits normal terminal screws.

A large, heavy-duty mechanic's screwdriver is needed for turning down terminals in the service entrance box.

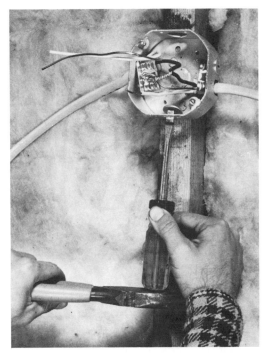

A screwdriver is also ideal for punching knockout holes in junction boxes.

suitable for electrical work. You'll need three different types to do most work properly. The first is naturally called an electrician's screwdriver. This is a long thin-shanked screwdriver with a medium- to small-size head and is used for reaching into tight places to turn tiny terminal screws.

The handles on these and all screwdrivers used for electrical work should be well-insulated plastic. But again don't count on the insulation to protect you. *Shut off the power.* A screwdriver can slip out of the slot, letting the metal blade contact two hot points and causing a bad short and a melted screwdriver blade. It's a good idea to get a screwdriver with an insulated blade shaft as well as an insulated handle.

Because of its long thin shank, an electrician's screwdriver can't be used to apply very much pressure. So you need a good stout, medium-shank screwdriver for loosening stubborn screws such as those on boxes. And although not a necessity, a huge long-bladed mechanic's screwdriver really makes one job easier: This just can't be beat for reaching into and applying the pressure needed to turn the large terminal screws on breaker and fuse boxes.

BRACE AND BITS

A good set of brace and bits becomes invaluable if you're doing any wiring work on an older building. Get a good ratchet brace so you can work it in corners or against a wall. You'll need a set of bits running from ¼ to 1 inch. For working on older buildings and adding wiring to existing work, you'll also need an extension for the bits. This allows you to bore the depths needed to reach through ceiling plates and floor joists, and into the walls.

YANKEE DRILL

Although not a necessity, these small push-type drill-screwdrivers are great for starting screw holes in paneling and wood studs.

PORTABLE ELECTRIC DRILL

If you own just one power tool, this is probably it. If the house you're working on has power supplied, a power drill can really save you a lot of work. Normally a power company will supply a temporary meter and fuse box hookup for power while you build a new house. You'll need a ¼-inch bit for cutting starter holes in lath-and-plaster or wallboard, as well as a 1-inch paddle bit for

An extra long bit like this one lets you make through-the-floor passageways. You can also use an extension bit that mounts ahead of a regular one.

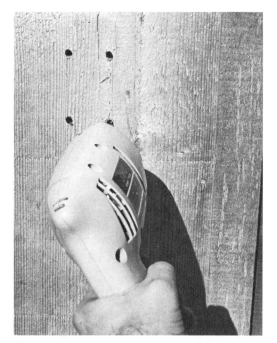

If you have power at the job site, a portable electric drill is a great help in making starter holes for your saw and passageways for wire runs.

A portable saber saw can speed up the many cuts for receptacles, switches and junction boxes.

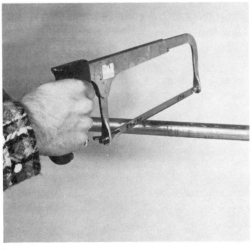

For conduit work, you'll need a hacksaw. Or you can use plumber's pipe cutters. All cuts must be reamed.

cutting wire-running holes through studs and joists. The paddle bits cut faster than almost any other bit in soft wood, although they will cause a great deal of splintering. A right-angle drive attachment for your portable electric drill can also be a great help in getting into tight spots.

SABER SAW

Along with a portable electric drill, about the only power tool equipment you might need is a saber saw. These are great for cutting receptacle, switch and ceiling fixture box holes in wallboard, lath-and-plaster or paneling.

KEYHOLE SAW

Or you can use a keyhole saw for cutting these holes. It's a little slower than the saber saw, but the results are the same, and if you don't have electricity on the site, a keyhole saw is a necessity.

HACKSAW

If you're doing new work, in some areas you'll be required to install it in conduit. Conduit piping can easily be cut with a hacksaw fitted with a fine-toothed blade. Don't use a coarse-toothed blade, or you'll only make the blade chatter, causing rough, uneven cuts that will be hard to fit and that could tear wiring insulation.

HAMMER

Almost any hammer you might have around the house will do for your electrical chores. You'll need it for driving cable and wire staples, for fastening boxes to studs and joists, and for knocking out knockouts in boxes.

WOOD CHISELS

If installing wiring in an older house,

you'll have to do a great deal of carpentry work. For this, you'll need a set of chisels for chiseling raceways and for chiseling around ceiling plates and doorways. Chisels can also be used for cutting notches into studs, when you install conduit in new work. A couple of chisels will suffice: A 1-inch chisel will notch new studs, and for work in an older house a thin-bladed chisel will cut the tongues off tongue-and-groove flooring so that you can lift the flooring and install wiring underneath.

TAPE RULE

A good tape rule is indispensable for measuring runs and figuring the amount of material needed. It's also necessary for measuring locations for switches, receptacles and boxes. A good wiring job will have all switches the same height and all receptacles the same height from floor and from countertops. The tape rule should extend at least 12 feet, and a second 25-foot rule is a good investment.

LEVEL

A good 3-foot level is a must for doing carpentry work that accompanies rewiring. It is especially useful for locating boxes correctly on studs and for setting boxes in masonry walls.

WRENCHES

If you're doing any conduit work you'll need a set of wrenches. A couple of adjustable wrenches will do just fine and will handle almost any situation you might encounter.

STUD FINDER

These inexpensive devices are great for locating studs in older buildings and thereby aid you in the installation of switches and

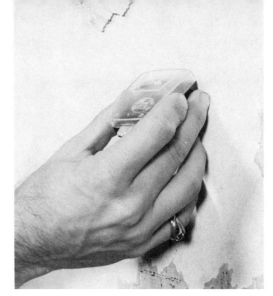

A stud finder contains a magnet on a swivel that points to hidden nails in studs.

receptacles. Housed inside the stud finder's plastic case is a magnet on a shaft. Whenever the stud finder is moved to within ¾ inch of a nail, the magnet swings toward it.

WAIST TOOLBAG

Although a toolbag is definitely not a necessity, after crawling through the attic on your belly or through a tight crawl space under the house, then finding you don't have the right tool, you'll wish you had one. They're made of leather and have pockets that carry everything from pliers to screwdrivers, and they have a small chain with a hook on the end for holding rolls of electrician's tape.

SOLDERING TOOLS

Again if you have electricity on the site, an electric soldering gun just can't be beat for making those tight solder connections. But if there is no electricity, you'll have to settle for the old-fashioned soldering irons heated with a propane torch. You'll want to check on your local wiring codes regarding soldering, because in some areas it is not

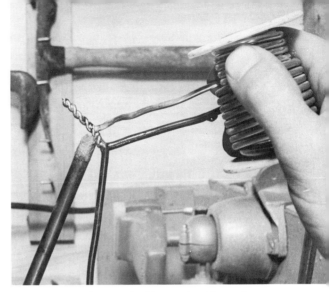

Before soldering, coat the splice with a noncorrosive flux, sold in paste form.

Then with a hot iron or electric gun, heat the splice until solder held against it flows into the twists.

required, and it may even be prohibited. In other areas, it is required for certain situations.

TESTERS

There are several testers, and we will discuss a couple in Chapter 16. The one tester you'll want before you even begin a job is the small two-prong tester that has a tiny light bulb in the end. This tester determines if a receptacle is hot, if the wires coming into a box are hot, or which wire is hot and which isn't when you are installing 3-way switches.

These little devices are quite inexpensive and can save you a really bad shock. On more than one occasion, I have turned off the power at the circuit breaker or fuse box for a circuit I was working on, then tested it with the tester and found it still hot. It's not unusual to encounter a circuit mismarked at the breaker box. Believe me, I want to know if I'm working on a hot circuit or not, so the little tester gets a real workout.

It's a good idea to occasionally "test" the

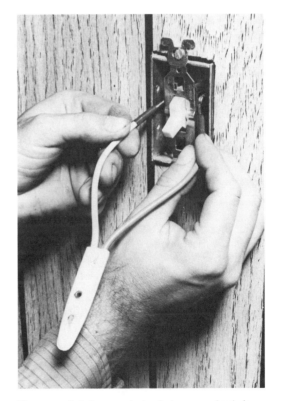

The essential 2-prong tester helps you check for a hot circuit that you think you're sure you turned off.

tester. Check it out with a receptacle that you know is working and make sure the tiny bulb is still functioning and that the small end probes haven't corroded over to the point that they don't make good contact.

CONDUIT BENDER

If you plan to work with thin-wall conduit piping, you'll need a conduit bender. Without this tool you will end up bending the soft metal pipe unevenly. The resulting kinks at each bend make it hard to pull wires through, and they can sometimes break, leaving jagged edges that tear insulation and cause shorts. A conduit bender is

For conduit you'll need a special conduit bender that makes a gradual, even bend without kinks that would otherwise catch or tear wire. First step is to fit the bender over the conduit. Second, with the conduit on the floor, pull the bender back until you feel resistance; and then step on the heel of the bender to start the bend. Third, keeping the conduit in place with your foot, complete the bend.

a sort of clamp with a rounded end, as shown in accompanying photos. It is slipped over the end of the conduit and slid into position. The pipe and bender are then positioned on the ground, with the bender handle up. Stepping on the edge of the bender provides the initial power to start the pipe bending, and from there on it's merely a matter of pulling back on the handle until the desired bend is achieved. Most codes will not allow more than a 90° bend in conduit, which is what most benders will do without any extra stress or effort. After the pipe is bent into the shape desired, the bender is merely slipped off the end or to the position of the next bend. The bender provides an even, smoothly bent curve without any kinks or tears.

Most thin-wall conduit comes from the manufacturer with inch marks that help you determine where to place the bender. For instance, if you wish to make a right-angle bend leaving 12 inches of conduit on the end to rise up from the floor, place the bender on the pipe with the arrow marked on the bender at 12 inches from the end of the conduit. (Note: Hand conduit benders are made for conduit in sizes from ½ inch to 1¼ inches, as well as for thin-wall and rigid conduit. You must be sure to match the bender to the size and rigidity of the tubing. However, for most household work you'll need only a ½-inch bender. The instructions accompanying conduit benders normally give exact methods of making various bends for that bender.)

PIPE CUTTER

A regular pipe cutter can also be used to cut thin-wall or rigid conduit, but it leaves more of an inside burr than does a hacksaw if the hacksaw is properly used, and cuts made by pipe cutters require more care as you ream out and "deburr" the end of the tubing.

PIPE REAMER

Any thin-wall conduit cut with either a hacksaw or pipe cutter must be reamed to insure there are no rough edges to catch and tear insulation. There are several types of reamers. A large pipe reamer used for ordi-

nary plumbing work can be used, although it is quite expensive. A much cheaper reamer, and one that will do the job as well, can be chucked into a brace. The reamer must be held perfectly square with the cut, and the edge must be reamed to absolute smoothness.

ARMORED CABLE CUTTERS

Some local codes require the use of BX, or armored cable—particularly for the installation of large motors, and in some cases for water heaters, furnaces and stoves. Although you can cut armored cable with a hacksaw, it's much easier to cut with a special tool made for that purpose. The cable-cutting tool resembles a pair of pliers and is used to snip, cut and pry away the cable. You can also use a good pair of aviation tin snips for this purpose.

The first step is to squeeze the cable tightly to make the interlocking, spiral strip buckle and leave an opening. Then twist the cable to unwind it an inch or so. Use the cable-cutting tool or tin snips to clip the armor strip, making sure that you don't cut or pinch any of the wires within. Using the

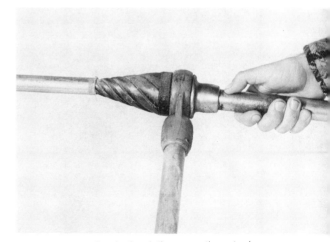

To prevent tears in wire insulation, ream the cut edge of the conduit, using a pipe reamer as shown or a less-expensive reamer that fits into a hand brace.

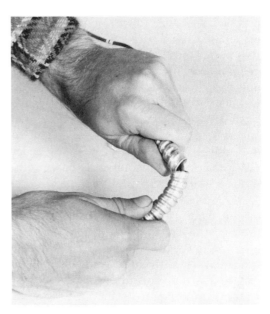

Local codes may require armored cable for some jobs. To cut it, first bend sharply until it parts, above left. Next, twist to further separate the continuous, spiral strip of metal.

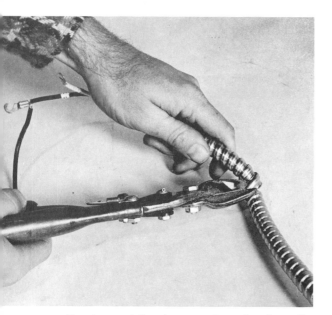

To cut, use aviation shears, as shown, tin snips, or the special cable shears in the next photo.

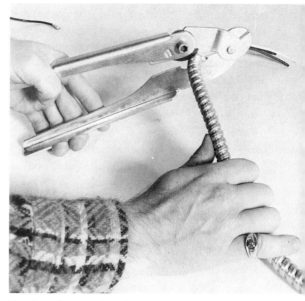

Use the pliers section of the cable shears or else a regular pliers to reshape the end of the cable.

cable pliers, you can reshape the end of the cable and slip it back in place, making a nice neat cut. The cut end of the cable must be secure, yet formed back in place to accept a fiber bushing which is pushed into the cable and over the protruding wires.

To cut armored cable with a hacksaw, hold the saw perpendicular to the armor strip and cut through only the one strip. Grasp the cable and twist to complete separating it. Then make a cut back far enough along the cable to reveal sufficient lengths of wires for the connection. Be sure not to cut into the wires. When you have completed cutting the armored sheathing, slide the end section of the sheathing off. Unwrap the paper from around the wires and jerk it out from inside the cable. Reshape the end of the cable and insert the fiber bushing.

FISH TAPE

Fish tape is a necessity when you work with conduit, as well as when you rewire an existing dwelling. The tape comes in many sizes and lengths. You can get it in a simple roll or wound on a spool fitted with a crank. The wire is extremely springy; so be careful when unrolling it. Or else you may get a slap in the face or, even worse, a gouged eye.

The tape is very thin and flexible and has a hook on each end. You push it through the conduit until it comes out the opposite end. Then you hook wires onto it and pull it back through. To help get the tape and wires through several bends, you can apply talcum powder. Fish tape is also helpful for running wires in older buildings. Tapes are run down from the attic or ceiling and fished out the receptacle or switch box openings.

Then the wires are pulled through. One hand-made fishing tool that also works well is a length of coathanger wire straightened out and hooked on each end. With this gadget you can fish wires that have been dropped down between studs in walls, as well as locate and fish out the end of a fish tape that invariably seems to be out of hand reach inside the wall, or in the ceiling.

Naturally work of this type is more easily done with two people, one pulling on the tape, and the other gently feeding the wires in place. The joint where the fish tape joins the wires should be made as small as possible, but secure, and should be wrapped tightly with tape to insure the wires don't catch on anything. You can also substitute a piece of ordinary galvanized steel wire for fish tape, but unless you're using the wire for short runs, you'll find it bends too easily and kinks in the run.

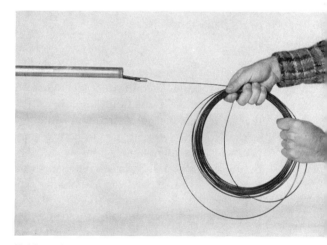

Fishing wire, consisting of spring steel with an end loop, pulls wire through conduit and passageways.

5

Materials

In the past 50 years wiring materials and methods have changed just as rapidly and drastically as the rest of the world's technology. Today's materials are much easier to use, and they're safer.

WIRES

Electricity is carried from one point to another through metal wires covered with various types of insulation. And the wires themselves are available in many different sizes and kinds. Among the kinds are copper wires and aluminum wires, which may be solid or stranded. Thin low-voltage wires handle small tasks, such as activating door bells. Heavy wires are used for circuit runs and service entrances. Small, stranded wires carry current in extension cords and light fixtures.

Copper has been the most common metal used in wire. Since problems with aluminum wire have caused many house fires, aluminum wires are not recommended for house circuits. Aluminum predominates in feeder lines and other long-distance trans-mission lines, primarily because of its lightness and strength.

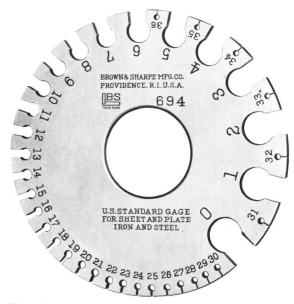

This wire and sheet metal gauge is shown actual size to depict relative sizes of electrical wires stripped of insulation. Sizes are determined in the straight, rather than the circular, portions of the slots.

In addition to wire, the basic materials for most jobs will include 1) junction boxes, 2) electrician's tape, 3) switch and receptacle boxes, 4) junction box covers, 5) receptacles, 6) switches, 7) twist-on connectors.

Wire Sizes

Wire size is determined by wire diameter in mils, expressed in circular mils. Since 1 mil equals $\frac{1}{1000}$th inch, a circular mil is the cross-sectional area of a circle with a 1 mil diameter, or $\frac{1}{1000}$th inch also.

In the United States, electrical wire is sized, or gauged, by a standard called the **American Wire Gauge**, often referred to as AWG. Household wires may range in size from hair-thin No. 60 up to the No. 0000 (usually expressed 4/0), nearly ½-inch thick. Thus the larger the wire, the smaller the number. One of the most commonly used household circuit wires, No. 14 measures about 64 mils in diameter, depending on type. The smaller sizes, such as No. 22, are used for low-voltage wiring, as for door bells.

Service entrance wires are normally from No. 6 to No. 2; although a service entrance to a 200-amp box would require up to a No. 000 (3/0) copper or a No. 0000 (4/0) aluminum. Stranded wire is gauged by the diameter of the overall wire, rather than the individual strands. The larger sizes of wire are normally available only in the stranded version.

Wire Uses

Although there are many different types of wires made, only a few are commonly used today in homes. All house wiring, except fixture wires (to 300 volts) and low-voltage wires, must be suitable for use up to 600 volts.

Wires are also limited to specific appli-

AMPACITIES

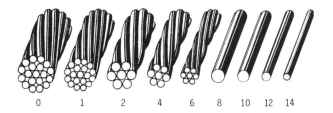

Wire size	In Conduit or Cable		In Free Air		Weather-proof Wire
	Type RHW* THW*	Type TW, R*	Type RHW* THW*	Type TW, R*	
14	15	15	20	20	30
12	20	20	25	25	40
10	30	30	40	40	55
8	45	40	65	55	70
6	65	55	95	80	100
4	85	70	125	105	130
3	100	80	145	120	150
2	115	95	170	140	175
1	130	110	195	165	205
0	150	125	230	195	235
00	175	145	265	225	275
000	200	165	310	260	320

* Types "RHW," "THW," "TW," and "R" are marked on the outer cover. Chart courtesy of Sears, Roebuck and Co.
• **R** type wire may be used in dry locations only and is no longer being manufactured.
• **TW,** the ordinary plastic insulated wire, may be used in wet or dry locations.
• **THW,** is similar to TW but will withstand a greater amount of heat; therefore, it has a higher ampacity rating in larger sizes.
• **RHW,** a rubber insulated wire with a protective braided covering, may be used either in wet or dry locations.

cations. Some wires may be used in damp locations, others in dry locations. In damp locations, use only those wires that are specified as approved for such areas. According to the Code, a damp location is an area somewhat protected—such as under a canopy or roofed porch, or in an average basement with a moderate degree of moisture. A dry location is not normally subject to dampness of any form, other than when the building is under construction. A wet location would be an underground location or any concrete or masonry location which is in constant contact with the earth; other wet locations subject to water saturation include washing areas and areas exposed to the weather.

Insulation

The most commonly used wiring today is made with thermoplastic insulation covering a copper or aluminum conductor. The thickness of the insulation depends on the size of the wire. The most common modern type is called TW and may be used in either wet or dry locations. There is also a THW which is similar, but it can withstand a greater degree of heat. (At one time rubber-covered wire was used extensively. But today you'll encounter it only in older homes.)

Other types of modern wire include THHN and THWN. Both of these are similar to the modern wires above, except that their insulation consists of a combination of thermoplastic covered with an extruded jacket of nylon. Because of the strength of the nylon insulator, these types of wires run a bit smaller and are normally used in conduit runs.

Another of the newer types of wire is XHHW, and this utilizes an even smaller type of insulator consisting of "cross-linked synthetic polymer." It has extraordinary insulation values and heat resistance and is becoming increasingly popular.

Cable

Two or more wires grouped together and covered by an additional outside insulation are called cable. Nonmetallic sheathed cable is widely used. A cable containing two No. 14 wires would be called 14-2. A cable carrying two No. 14 wires plus a ground wire would be called 14-2 *with ground*. If a cable carried three No. 14 wires, it would be called 14-3. The same system is used with other wire sizes such as No. 12; this, for example, could be 12-2 *with ground*. If a cable

carries two wires, one will always be white, the other black. A cable carrying three wires will normally contain a white, a black, and a red. The grounding wire may either be bare, green-insulated, or green with one or more yellow stripes. In a circuit the wires could run like this:

Circuit with 2 wires: white, black.
   with 3 wires: white, black and red.
   with 4 wires: white, black, red and blue.
   with 5 wires: white, black, red, blue and yellow.

Because cable is made with up to 3 wires *plus* ground, any circuits carrying more than 4 wires, including ground, must be run in conduit.

There are basically two types of cable, type NM and type NMC. Type NM is commonly called Romex or Loomwire and consists of individual wires covered with insulation. Each wire is also spirally wrapped with paper. Over the entire bundle of wires is an outer wrapping. In the older wires a fibrous material was used, but in the newer cable the outer sheathing is plastic.

Type NMC is called a dual purpose cable and is often used in all types of wiring, dry

There are two basic types of plastic-sheathed cable. NM consists of paper-wrapped wires and an outer sheath. NMC can be used for wet or dry wiring, and its individual wires are embedded in plastic.

RATINGS OF CONDUCTORS and TABLES to determine VOLT LOSS

With higher ratings on new insulations, it is extremely important to bear VOLT LOSS in mind, otherwise some very unsatisfactory experiences are likely to be encountered.

These tables take into consideration REACTANCE ON AC CIRCUITS as well as resistance of the wire.

Remember on short runs to check to see that the size and type of wire indicated has sufficient ampere capacity.

How to figure volt loss

MULTIPLY DISTANCE (length in feet of one wire) by the CURRENT (expressed in amperes)

by the FIGURE shown in table for the kind of current and the size of wire to be used.

THEN put a decimal point in front of the last 6 figures AND — you have the VOLT LOSS to be expected on that circuit.

Example — No. 6 copper wire in 180 feet of iron conduit — 3 phase, 40 amp. load at 80% power factor.

Multiply feet by amperes: 180 x 40 = 7200

Multiply this number by number from table for No. 6 wire three-phase at 80% power factor:

7200 x 735 = 5292000

Place decimal point 6 places to left.

This gives volt loss to be expected: 5.292 volts

(For a 240 volt circuit the % voltage drop is $\frac{5.292}{240}$ x 100 or 2.21%.)

How to select size of wire

MULTIPLY DISTANCE (length in feet of one wire) by the CURRENT (expressed in amperes)

DIVIDE that figure INTO the permissible VOLT LOSS multiplied by 1,000,000

Look under the column applying to the type of current and power factor for the figure nearest, but not above your result

AND — you have the size of wire needed.

Example — Copper wire in 180 feet of iron conduit — 3 phase, 40 amp. load at 80% power factor — volt loss from local code equals 5.5 volts.

Multiply feet by amperes: 180 x 40 = 7200

Divide permissible volt loss multiplied by 1,000,000 by this number:

$\frac{5.5 \times 1,000,000}{7200} = 764$

Select number from table, three-phase at 80% power factor, that is nearest but not greater than 764. This number is 735 which indicates the size of wire needed: No. 6

COPPER CONDUCTORS

| | | AMPERE RATING | | | | VOLT LOSS (See Explanation Above) | | | | | | | | | |
| | WIRE SIZE | Type T, TW (60°C Wire) | Type RH, RHW, THW THWN, THHN (75°C Wire) | Type RHH, RHHW XHHW (90°C Wire) | Direct Current | THREE PHASE — 60 Cycle, lagging power factor | | | | | SINGLE PHASE — 60 Cycle, lagging power factor | | | | |
						100%	90%	80%	70%	60%	100%	90%	80%	70%	60%
COPPER CONDUCTORS IN IRON CONDUIT	14	15	15	15	6100	5280	4800	4300	3780	3260	6100	5551	4964	4370	3772
	12	20	20	20	3828	3320	3030	2720	2400	2080	3828	3502	3138	2773	2404
	10	30	30	30	2404	2080	1921	1733	1540	1340	2404	2221	2003	1779	1547
	8	40	45	50	1520	1316	1234	1120	1000	880	1520	1426	1295	1159	1017
	6	55	65	70	970	840	802	735	665	590	970	926	850	769	682
	4	70	85	90	614	531	530	487	445	400	614	613	562	514	462
	3	80	100	105	484	420	425	398	368	334	484	491	460	425	385
	2	95	115	120	382	331	339	322	300	274	382	392	372	346	317
	1	110	130	140	306	265	280	270	254	236	306	323	312	294	273
	0	125	150	155	241	208	229	224	214	202	241	265	259	247	233
	00	145	175	185	192	166	190	188	181	173	192	219	217	209	199
	000	165	200	210	152	132	157	158	155	150	152	181	183	179	173
	0000	195	230	235	121	105	131	135	134	132	121	151	156	155	152
	250M	215	255	270	102	89	118	123	125	123	103	136	142	144	142
	300M	240	285	300	85	74	104	111	112	113	86	120	128	130	131
	350M	260	310	325	73	63	94	101	105	106	73	108	117	121	122
	400M	280	335	360	64	55	87	95	98	100	64	100	110	113	116
	500M	320	380	405	51	45	76	85	90	92	52	88	98	104	106
	600M	355	420	455	43	38	69	79	85	87	44	80	91	98	101
	700M	385	460	490	36	33	64	74	80	84	38	74	86	92	97
	750M	400	475	500	34	31	62	72	79	82	36	72	83	91	95
	800M	410	490	515	32	29	61	71	76	81	33	70	82	88	93
	900M	435	520	555	28	26	57	68	74	78	30	66	78	85	90
	1000M	455	545	585	26	23	55	66	72	76	27	63	76	83	88
COPPER CONDUCTORS IN NON-MAGNETIC CONDUIT Lead covered cables or installation in fibre or other non-magnetic conduit, etc.	14	15	15	15	6100	5280	4790	4280	3760	3240	6100	5530	4936	4336	3734
	12	20	20	20	3828	3320	3020	2700	2380	2055	3828	3483	3112	2742	2369
	10	30	30	30	2404	2080	1910	1713	1513	1311	2404	2202	1978	1748	1512
	8	40	45	50	1520	1316	1220	1100	976	851	1520	1406	1268	1128	982
	6	55	65	70	970	840	787	715	641	562	970	908	825	740	648
	4	70	85	90	614	531	517	466	422	374	614	596	538	486	431
	3	80	100	105	484	420	410	379	344	308	484	474	438	397	355
	2	95	115	120	382	331	326	303	278	250	382	376	350	321	288
	1	110	130	140	306	265	266	251	232	211	306	307	289	267	243
	0	125	150	155	241	208	216	206	192	176	241	249	237	221	203
	00	145	175	185	192	166	176	170	160	148	192	203	196	184	171
	000	165	200	210	152	132	145	141	134	126	152	167	163	155	145
	0000	195	230	235	121	105	119	118	114	108	121	137	136	131	125
	250M	215	255	270	102	89	105	106	104	100	103	121	122	120	115
	300M	240	285	300	85	74	92	95	93	91	86	106	109	107	105
	350M	260	310	325	73	63	82	85	84	83	73	94	98	97	96
	400M	280	335	360	64	55	75	78	79	78	64	86	90	91	90
	500M	320	380	405	51	45	64	69	71	70	52	74	80	82	81
	700M	355	420	455	43	38	57	63	66	66	44	66	73	76	76
	600M	385	460	490	36	33	53	58	61	63	38	61	67	70	73
	750M	400	475	500	34	31	51	56	60	61	36	59	65	69	70
	800M	410	490	515	32	29	49	55	58	60	33	57	64	67	69
	900M	435	520	555	28	26	46	52	55	57	30	53	60	64	66
	1000M	455	545	585	26	23	43	50	54	56	27	50	58	62	64

or wet or corrosive. NMC type cable normally consists of the individual wires embedded in solid plastic. Make sure the cable is stamped "NMC." This type of cable is normally available only with the ground wire bare, or uninsulated.

Another type is armored cable, sometimes called by its trade name BX. This consists of a spirally wound steel flexible cable with the individual insulated wires running through it. Around the wires is a layer of paper, which acts as protection against abrasion from the steel sheathing. An internal bonding strip wire normally runs inside the armor and outside the paper sheathing.

Due to the high cost of copper, aluminum is being used a great deal for service-entrance cables and feeders for heavy motors and other appliances. It should not, however, be used for branch circuits in homes. A copper-clad aluminum cable is considered safe, but the savings with this type of cable are negligible.

ALUMINUM CONDUCTORS

	WIRE SIZE	AMPERE RATING			Direct Current	VOLT LOSS See Explanation Above									
		Type T, TW (60°C Wire)	Type RH, THWN RHW, THW (75°C Wire)	Type RHH, THHN XHHW (90°C Wire)		THREE PHASE — 60 Cycle, lagging power factor					SINGLE PHASE — 60 Cycle, lagging power factor				
						100%	90%	80%	70%	60%	100%	90%	80%	70%	60%
ALUMINUM CONDUCTORS IN IRON CONDUIT Size that can be used with FUSETRON D.E. fuse or LOW-PEAK D.E. fuse	12	15	15	15	6040	5230	4760	4260	3740	3243	6040	5500	4920	4320	3745
	10	25	25	25	3800	3291	3005	2690	2380	2080	3800	3470	3110	2750	2395
	8	30	40	40	2390	2070	1905	1725	1525	1330	2390	2200	1990	1760	1540
	6	40	50	55	1530	1325	1238	1126	1005	890	1530	1430	1300	1160	1030
	4	55	65	70	966	837	795	726	647	585	966	918	838	747	675
	2	75	90	95	606	526	511	473	434	397	606	590	546	498	456
	1	85	100	110	480	415	414	386	355	330	480	478	446	410	380
	0	100	120	125	382	331	336	317	294	277	382	388	366	340	320
	00	115	135	145	302	262	272	260	244	232	302	314	300	282	268
	000	130	155	165	240	210	225	217	206	199	242	260	250	238	230
	0000	155	180	185	192	168	185	182	175	173	194	214	210	202	200
	250M	170	205	215	161	142	163	163	157	153	164	188	188	181	177
	300M	190	230	240	134	119	141	142	140	141	137	163	164	162	163
	350M	210	250	260	115	102	126	128	127	125	118	146	148	147	148
	400M	225	270	290	101	91	115	120	119	122	105	133	138	137	141
	500M	260	310	330	80	74	100	104	106	107	85	115	120	122	124
	600M	285	340	370	67	62	88	95	98	101	72	102	110	113	117
	700M	310	375	395	58	55	82	88	92	97	64	95	102	106	112
	750M	320	385	405	54	52	79	85	89	94	60	91	98	103	108
	800M	330	395	415	50	49	76	83	87	93	57	88	96	101	107
	900M	355	425	455	45	45	72	80	83	88	52	83	92	96	102
	1000M	375	445	480	40	42	68	76	81	85	48	79	88	93	98
ALUMINUM CONDUCTORS IN NON-MAGNETIC CONDUIT Lead covered cables or installation in fibre or other non-magnetic conduit, etc.	12	15	15	15	6040	5230	4750	4250	3720	3217	6040	5490	4900	4300	3715
	10	25	25	25	3800	3290	3000	2680	2360	2040	3800	3460	3100	2730	2360
	8	30	40	40	2390	2070	1900	1701	1501	1304	2390	2190	1970	1740	1510
	6	40	50	55	1530	1325	1230	1110	990	866	1530	1420	1280	1140	1000
	4	55	65	70	966	837	787	715	641	570	966	908	826	740	656
	2	75	90	95	606	525	504	462	419	378	606	580	534	484	435
	1	85	100	110	480	416	405	376	343	312	480	468	434	396	360
	0	100	120	125	382	331	328	307	282	258	382	378	354	326	299
	00	115	135	145	302	262	265	251	232	216	302	306	290	268	249
	000	130	155	165	240	208	217	206	175	164	240	250	238	202	189
	0000	155	180	185	192	166	177	171	161	154	192	204	197	186	176
	250M	170	205	215	161	139	153	151	144	138	161	177	174	166	159
	300M	190	230	240	134	116	133	132	127	125	134	153	152	147	144
	350M	210	250	260	115	100	117	117	114	114	115	135	135	132	131
	400M	225	270	290	101	87	106	106	106	105	101	122	124	122	121
	500M	260	310	330	80	70	89	92	92	91	81	103	106	106	105
	600M	285	340	370	67	59	79	83	83	83	68	91	96	96	96
	700M	310	375	395	58	50	71	76	78	82	58	82	88	90	94
	750M	320	385	405	54	48	68	73	75	76	55	79	84	87	88
	800M	330	395	415	50	44	66	71	73	74	51	76	82	84	86
	900M	355	425	455	45	40	61	67	69	71	46	70	77	80	82
	1000M	375	445	480	40	36	57	63	66	67	41	66	73	76	78

These charts are reprinted with permission of Bussman Mfg., Div. of McGraw-Edison Co. (© revised 1974).

This is a 2-wire service drop cable. One wire is rubber covered, the other twisted aluminum.

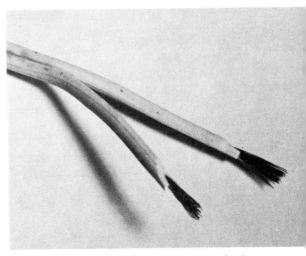

Stranded wire is commonly used as lamp and extension cord material.

Armored cable consists of a flexible spiral-wound metal sheathing that is prewired.

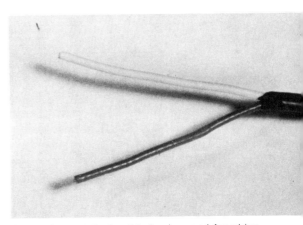

Called thermostat wire, this is also used for wiring door chimes and even speaker runs.

Other Wiring

Other types of wiring include flexible cords such as those used on lamps and portable appliances. These are normally made of stranded wires embedded in a solid rubber insulation, and are called type SP. If the insulation is plastic, the wire is called SPT.

There are several other types of wires used in house wiring, but one of the most common is low-voltage wire. Because of the low voltage traveling through the wires, normally under 30 volts, there is little insulation. Another similar wire is thermostat cable, which consists of two or more separate wires encased in a tiny rubber or plastic cable. Because it's so much easier to run the two wires of a circuit at the same time, this type of "cable" is most commonly used in wiring low-voltage items like those for doorbells.

The main idea in selecting wire is to select the proper size and type of wire to do the job. The wire should be large enough in diameter to carry the amperage needed without overheating. Even though a wire may be large enough to carry the amperage, the voltage drop caused by the length of the wire may also give trouble. The tables on pages 44 and 45 illustrate how to calculate the voltage drop in a given size wire for a specific length. Another most important consideration in selecting wires is that they must conform to local codes.

CONDUIT

So far we have discussed the various types of wires including plastic-sheathed cable and armored cable. In some instances the individual wires will be carried in runs of metal tubing called thin-wall tubing or EMT. Many local codes may require EMT for new buildings and additions.

Thin-walled conduit (E.M.T.) may be required by local codes. It comes in 10-foot lengths and bears the label of the Underwriters Laboratories Inc.

There are basically five kinds of conduit: rigid metal conduit and intermediate metal conduit (IMC) which are made up of a steel pipe similar to water pipe, thin-wall conduit referred to as EMT, nonmetallic (plastic) conduit, and flexible metal conduit. Most single-home wiring installations of conduit use thin-wall conduit. It's cheaper and easier to work, and it requires fewer tools. Also known as electrical metallic tubing, EMT conduit comes in 10-foot lengths and has the Underwriters' label on it. It is available in sizes up to and including 4 inch, and should be fitted together with various types of joint connectors and special threadless fittings. (Don't try to thread lengths of EMT together.) The special fittings hold the tubing together with pressure.

Flexible conduit is very similar to armored cable, except there are no wires in it. The wires are pulled through after the conduit is installed.

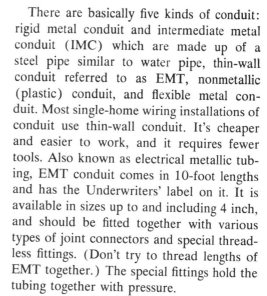

These are thin-wall conduit connectors and fittings.

METAL BOXES

PLASTIC BOXES

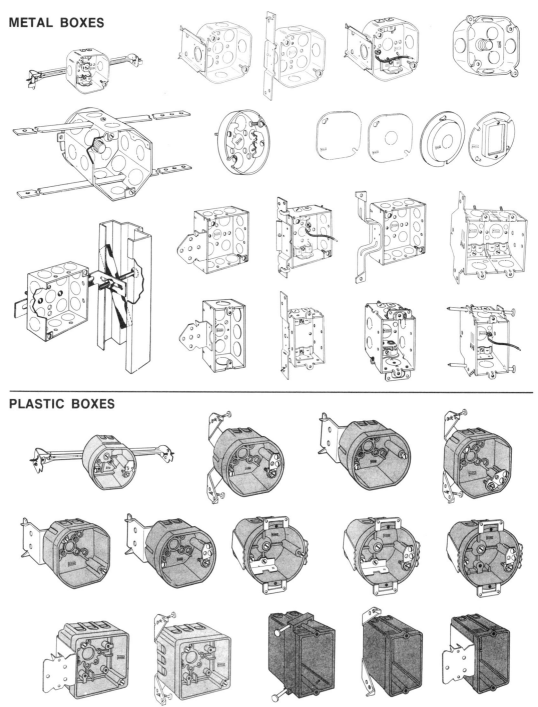

This shows some of the more common metal boxes, plastic boxes, and their fittings. Courtesy of Raco.

BOXES

There are literally hundreds of kinds and sizes of boxes for use in electrical work. The boxes most commonly used in house wiring are junction boxes, switch boxes, light fixture boxes, and receptacle boxes. Boxes are available in either metal or plastic. Before using plastic boxes, check with your local codes as to whether they are acceptable and, if so, in what types of situations.

The Code requires that boxes be used in any location where wires are spliced, connected, or connected to terminals of any type of electrical equipment. The boxes must also be properly supported in reference to the various types of installations and the boxes.

The most common junction or fixture outlet boxes may be either octagonal or square, and they're available in many different sizes. On some boxes knockout holes are provided at different locations. You pry out the prepunched hole and insert a cable clamp to hold the cable in place. Other boxes have a cable clamp in their bottoms, and you merely screw it in place to secure the cable. In addition to the octagonal and square boxes, there are also rectangular boxes for switches and receptacles.

The most common box depth is 2½ inches, although some are as deep as 3½ and some only 1½ inches. The thinner boxes may be used only in areas where deeper boxes might damage the building structure. The chapter on basic techniques describes how to install and use the various types of boxes.

Outlet boxes may be supported on special

In some boxes, you remove the knockouts by twisting a screwdriver blade in knockout slots.

Some metal boxes allow your joining them to make a larger box. Here, remove the side plates and simply screw the boxes together.

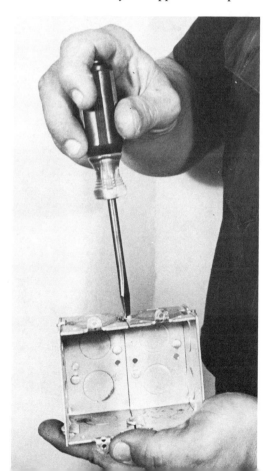

hanger bars which are factory installed on the boxes, or on special bars which you can fasten to the boxes. Some boxes are for outdoor use and are specified as waterproof. In addition to the boxes, there are covers. Each junction box containing spliced wires must be covered by a suitable cover. Each receptacle and switch box must also be covered by a cover appropriate for the item used in the box.

The number of wires in a box is limited according to the code rules. Charts issued by box manufacturers indicate the number of wires that may be used in various sizes, both in metal and in plastic boxes.

RECEPTACLES

The Code requires that all receptacles be grounded except for a replacement receptacle on an existing system where a grounding means does not exist. This also means that receptacles must have the silver colored and brass colored terminals, and a green grounding terminal as well. Receptacles today come with receptacle plates in all the colors of the rainbow to match various decors.

Most modern receptacles are also fitted with "push-in" connector holes in their back, so you no longer have to bend the wire to fit it around the terminal. Merely

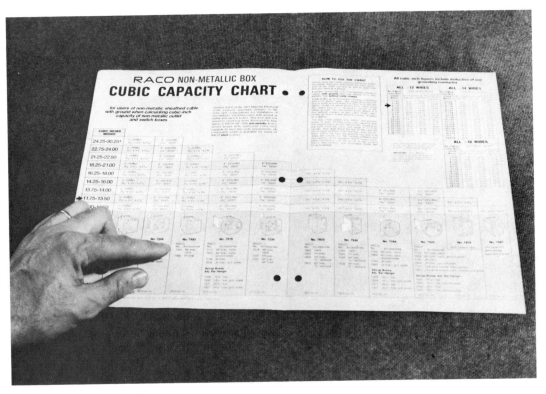

Box capacity charts like this one help you determine the proper size boxes for the many possible wiring combinations for switches and receptacles.

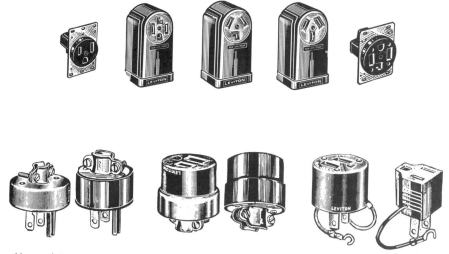

Heavy-duty receptacles are designed to receive the grounded, 3-prong plug. Courtesy of Leviton Industries.

pushing the wire in place connects it to the receptacle.

SWITCHES

Again, there are as many different styles, colors and kinds of switches as there are receptacles. Picking the right switch for the right job is very important. Make sure the switch you select is rated for the job. You'll find the rating stamped on the switch. Most switches are rated at 10 amps, 125 volts; or 5 amps, 250 volts. This means the switch can control 10 amps if the voltage is not higher than 125 volts, or 5 amps if the voltage is not higher than 250. One way of making sure you have the proper switch, when replacing an old one, is to take the old switch with you when buying.

The most commonly used switch is a simple toggle switch. This uses a simple L-shaped armature which is operated by an "up-and-down" lever to contact the terminals. Newer toggle switches are "silent" and don't make the clicking sound; however, they operate on the same principle.

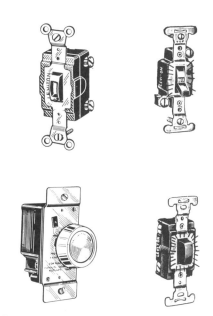

Common switches like these can be activated by flipping a lever, pushing a button, or turning a dial that determines light intensity. Special switches are available that respond to timer mechanisms and photocells. Courtesy of Leviton Industries.

There are mercury switches which are also silent. These contain a small cylinder of mercury which pivots, creating the contact between the terminals.

The newest type of switch is the "touch switch." It comes in all decorator colors and is activated by merely pushing in the top or bottom portion of the switch, or on the older units, by merely pushing in the center contact button. A small ratchet operates the switch to turn the switch to the terminals.

Some switches are equipped with small lights which enable you to see the switch when you enter a dark room.

Today, dimmer switches are among the types for dining rooms and living rooms. These use a rheostat that enables you to turn the switch-dial down to dim the light and up for full bright.

FIXTURES

As with switches, there are many styles of fixtures, but they can be grouped into several basic types. The first type is a simple ceiling light. The second includes wall lights mounted, perhaps on a bathroom wall. Most are controlled by a switch or switches mounted on the wall. There are ceiling fixtures that provide central lighting to a room, as well as ceilng fixtures mounted with low-wattage bulbs that merely add atmosphere. There are also ceiling-mounted fluorescent fixtures mounted above a recessed ceiling and covered with a plastic grid that throws an even glow over an entire area.

In addition there are recessed lights which may be used as accent lights around a room or to throw a tiny "spot" of light on a conversation piece or painting. All are controlled by a switch or switches mounted on the wall. In the bathroom you may wish to install a combination, light-heater-fan unit which warms you as you step from the tub.

1-GANG SWITCH

2-GANG SWITCH

3-GANG SWITCH

4-GANG SWITCH

5-GANG SWITCH

6-GANG SWITCH

1-GANG PUSH SWITCH

2-GANG PUSH SWITCH

1-GANG DUPLEX RECEPTACLE

2-GANG DUPLEX RECEPTACLE

1-GANG SINGLE RECEPTACLE

2-GANG SINGLE RECEPTACLE

2-GANG, 1 SWITCH AND 1 SINGLE RECEPTACLE

2-GANG, 1 SWITCH AND 1 DUPLEX RECEPTACLE

2-GANG, 1 SINGLE AND 1 DUPLEX RECEPTACLE

3-GANG, 2 SWITCHES AND 1 DUPLEX RECEPTACLE

1-GANG BLANK

2-GANG BLANK

3-GANG BLANK

1-GANG TELEPHONE

1-GANG TELEPHONE, SQUARE RECEPTACLE

1-GANG LOUVRE

Direct-lighting fixtures come in a variety of styles from practical to ornamental, with choices among fluorescent and incandescent lights, authentic and simulated materials.

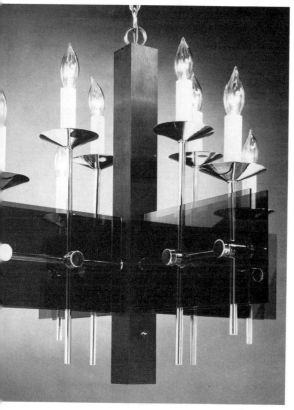

SERVICE ENTRANCE EQUIPMENT

The basic equipment in the service entrance is the panel or service box. This is a metal box which encases the electrical manual power disconnect device, as well as the overcurrent protection devices. These come in various sizes depending on the amperage needed for the service. The boxes are available with either a circuit-breaker type of overcurrent protection or a fuse type of protection. The main fuse or pullout fuse block contains the main fuses. Pulling out the fuse block shuts off the electricity to the circuits serviced by the electrical power, *but not to the panel itself.* Use all safety precautions in handling any type of service panel.

The individual fuses for the circuits may range anywhere from 12 to 50 amps depending on the circuit load and are round with a base like a light bulb which is screwed into the fuse panel. The top of the fuse is clear glass or plastic and allows you to see the tiny strip of metal encased in the fuse. In the event of a short circuit or overloaded circuit, the fuse is designed so that the tiny metal strip will melt and separate before the circuit becomes overheated,

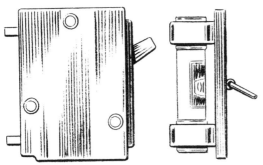

These drawings show a side view of a typical circuit breaker, left, and a fuse block with a pullout fuse.

This is the long-established circuit fuse that screws into a standard socket in the service panel.

This assembly prevents "overfusing." The adapter on the left screws into the fuse panel and locks in place. It will then accept only fuses of one amperage.

thereby shutting off the circuit. Determine the correct size amperage fuse for the circuit, as described in Chapter 2.

A circuit breaker, on the other hand, looks like the one on page 54. Some contain a carefully calibrated metal strip somewhat like that used in a thermostat. If the current causes the strip to overheat, the strip bends and releases a trip that interrupts the circuit, shutting it off. A more commonly used type is the magnetic circuit breaker which employs a solenoid.

MISCELLANEOUS ITEMS

In addition to the above, there are also many other items used for electrical wiring. The first and most important are the connectors. Most of today's wiring is done with twist-on connectors—in areas where local codes permit. These are plastic shells containing metal threads. When a shell is twisted over the wires, it brings them all in contact. To disconnect the wires, you merely remove the connectors. It is vital that you use the proper size connectors for the sizes and numbers of wires used in them. Too large a connector will not make a proper connection, and too small a connector will allow a portion of the wires to protrude with the possibility of a short.

Here, wires are being connected to rafter-mounted porcelain service drop insulators.

Twist-on connectors make for easy hookups and provide reliable insulation of wire ends.

This is a tension-type service drop holder.

Surface-Mounted Devices

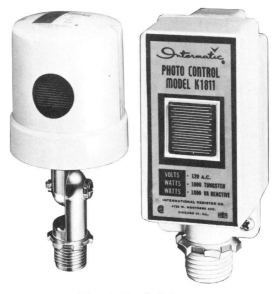

Electric Eye Switches

Many miscellaneous items can make night life safer
and more convenient.

**Outdoor
Receptacle
Cover Plate**

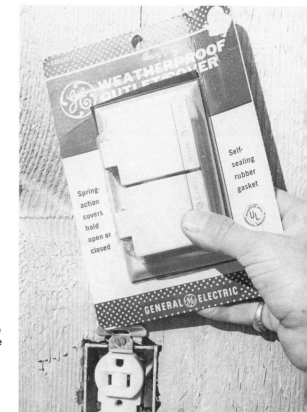

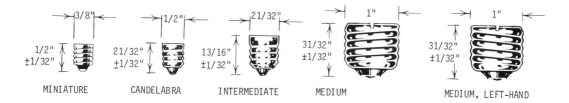

MINIATURE   CANDELABRA   INTERMEDIATE   MEDIUM   MEDIUM, LEFT-HAND

In addition to these new types of connectors, there are also solderless clamp connectors which are merely open-end plastic tubes lined with metal. The two stripped wires are inserted in the plastic shell and the ends crimped to fasten them together. Some local codes don't allow the use of this type of connection, so check before using.

In addition there are hundreds of miscellaneous items including insulators for securing service drop lines, cable straps, surface

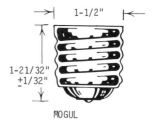

MOGUL

These are the standard screw-type lamp bases.

mounted receptacles and special outdoor lighting equipment.

Last, but not least, is the old familiar electrician's tape. At one time this was made of composition, but today's tape is normally plastic. It is used for everything from wrapping joints which have been soldered, to making a "permanent" insulated connection, to temporarily wrapping a wire before a circuit is completed. Buy plenty of this if you plan to do much wiring. You'll find a thousand uses for it.

A must for any wiring job, electrician's tape is useful both as insulation and fastener.

6

Basic Wiring

The "mechanics" of house wiring is actually quite simple. With the proper tools, materials and a little practice, the average do-it-yourselfer can handle most house wiring jobs. Like the chain that is only as strong as the weakest link, electrical wiring is only as good as its weakest connection, which could short and even cause a fire. Learn to use the proper materials and methods, and be methodical and careful in all jobs to insure the tasks are done properly and with good workmanship. You'll end up with a neater looking job and a safer system as well.

STRIPPING WIRE

For most ordinary house wiring, there are two methods and two tools that can be used for stripping the insulation from wires.

The easiest method for the beginning electrician employs wire strippers. Here you need a stripper that is sized for the most commonly used house circuit wires such as No. 14 and No. 12. These strippers may be designed like a pair of pliers with various jaw holes the size of the wires without their insulation. Or a stripper may be a "cutter" type, with a locking device that can be set for the various wire sizes. In any case, the best method is to clamp the strippers down over the insulation on the wire, quickly turn them around the wire to insure cutting through all sides of the insulation, then slide off the loosened piece of insulation. With a little practice, you'll be able to do this in seconds. The amount of wire to be stripped depends on what the wire will be connected to or into; this will be discussed later in the chapter.

Naturally, if the wire is as large as that used for entrance cable, you won't be able to strip it using ordinary wire strippers. In this instance use a pocketknife blade. The best blade is straight with a blunt end. Cut through the insulation, holding the knife at an angle so you don't nick the conductors. Then merely slice the insulation off. However, when stripping plastic-sheathed cable, pull the knife toward you along the length of the cable, with only the point of the blade slicing into the plastic. Be careful not to cut into the insulation of the individual wires. With the outer sheath of plastic sliced, peel

HOW TO STRIP WIRE AND CABLE

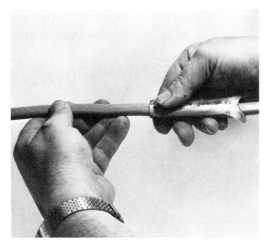

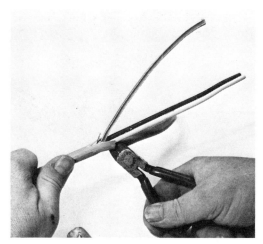

Strip Type NM cable with a cable stripper, as shown left. Then peel back the plastic sheathing and paper insulation.

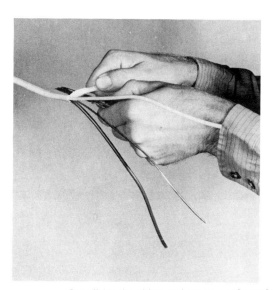

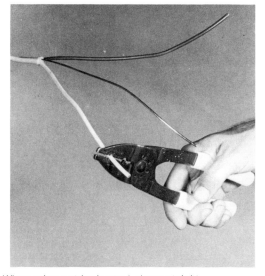

Cut off the sheathing and paper, as shown left. When using a stripping tool, shown at right, remove insulation by placing the wire in the proper jaw niche, spinning the tool, and slipping the end off.

it back and cut it off at the beginning of the slice. Then peel back the paper insulation and cut it off as well. The main idea in stripping wires is to remove all insulation to insure a good electrical connection, yet you should avoid making nicks or cuts above the stripped area, which could cause "shorts." When removing insulation from stranded

wires, be very careful that you don't slice through the individual strands, and weaken the wire.

JOINING WIRES

There are many different methods of joining or splicing wires. One of the simplest methods involves twisting them together

PROPER METHOD

WRONG METHOD

When stripping solid-core wire with a blade, you'll probably find it best to shave away from you.

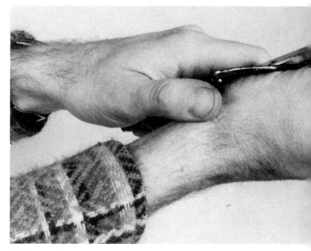

With small-stranded cable such as lamp cords, carefully slice off insulation, as shown, without knicking the strands.

using one of several different kinds of "splices." These splices must be mechanically strong to insure a good electrical connection. After splicing the wires, you can make a stronger mechanical connection by soldering the splice. Check with your local codes regarding spliced and soldered wires. Some local codes will not allow soldering.

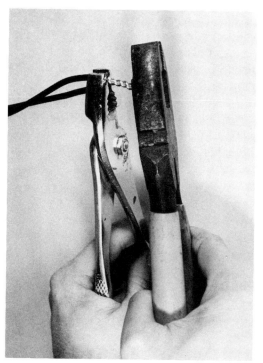

The pigtail splice must be strong enough to withstand stress and tight enough to conduct electricity properly. For this, you'll need two pairs of pliers.

Here you see the straight splice on the top and the pigtail splice on the bottom.

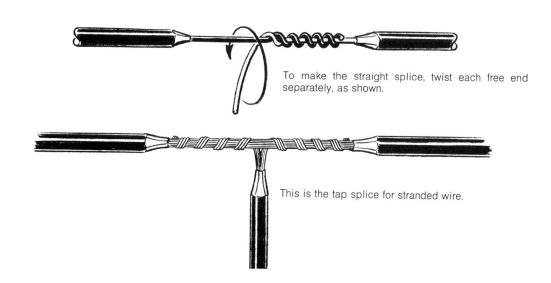

To make the straight splice, twist each free end separately, as shown.

This is the tap splice for stranded wire.

SOLDERING

Proper soldering of a wire joint takes a bit of practice. To begin, you should have clean conductors, the proper type and amount of flux and solder, and a clean tip on your soldering gun or hot iron. The first step in a soldering job is to make sure the copper wires are thoroughly cleaned of insulating material and show no traces of oil or dirt. Cleaning is essential, or the solder will not adhere to the metal. Don't attempt to solder aluminum; that's a job for the pros requiring special equipment and lots of practice.

To enable the solder to form a permanent bond with the copper, you'll need a flux of some sort. Don't use acid flux on electrical soldering because it can react with the cop-

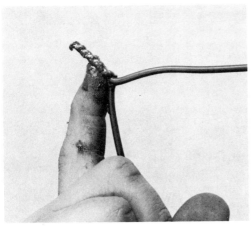

If your local code allows, you can solder the joints to make a stronger and more electrically efficient bond. In the method shown, a combination flux-and-solder paste is first rubbed onto the joint.

Here a torch is used to heat the wires quickly, so that the solder melts into the joint before the insulation becomes overheated. (Techniques for soldering with an electric gun are described in Chapter 4, "Tools.")

The result is a properly soldered splice.

CROSS
SECTION

Solder doesn't fill
between wires

An improper soldering job. Here the wire was not hot enough to allow the flow of solder into every opening.

per and form a sort of insulation, or the chemical reaction may eat away the conductors. The best flux is rosin, and a rosin-core solder makes the job even easier. In some cases by using rosin-core solder, you can skip the "fluxing" step. With the copper wires joined and cleaned, the next step is to heat the copper wires with the tip of a soldering gun or soldering iron or with a torch. The trick is to heat the wires themselves to a high enough temperature so that when the solder is touched to them it will melt and run into and around the joint. If the wires are not hot enough, the solder may just cover the joint and not entirely fill the spaces between the wires, thus making a poor connection.

On the other hand, it is sometimes quite tricky to heat the copper wires hot enough without melting the nearby insulation material. Make sure you don't accidentally touch the insulation on the wires or cause it to melt and leave an open space which could later cause a short. The secret of good soldering is to have the soldering gun or iron hot enough to heat the copper conductors very quickly. Then apply the solder immediately and get the job over before the heat has a chance to travel along the conductors and ruin the insulation. Once again, proper soldering takes a bit of practice, and it's a good idea to make a few trial runs on scrap pieces of wire before attempting a real splice. Soldering stranded wires is even trickier, because the strands may separate. Sometimes it's a good idea to apply a bit of solder to the ends of the strands of the individual wires before soldering them together.

TAPING

After splicing and/or soldering the joint, you must insulate it by taping. The Code requires that the joint be as well insulated as the continuous wires coming into the joint.

TAPING SPLICES

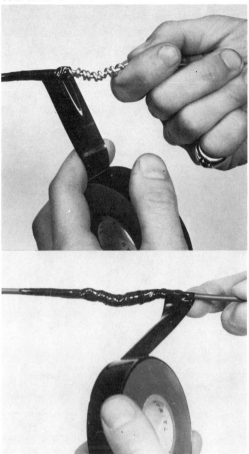

When taping a straight splice, hold the tape taut without stretching it. Overlap each wrap by half. Then make successive return wraps until the tape is built up as thick as the regular insulation.

To tape a pigtail splice, go around and between the insulated portions as well as the stripped ends.

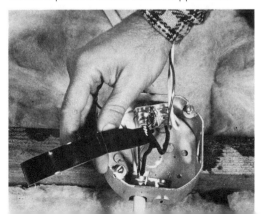

TEMPORARY END WRAP

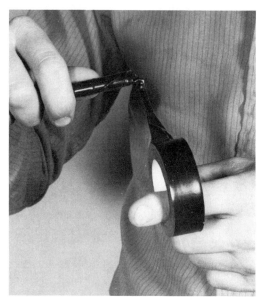

If you must leave a job before completing it, wrap the ends of the wires for safety. Begin as shown, left. When you reach the end of the cable, twist the roll around your finger to close off the opening.

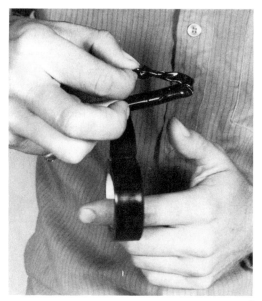

Leaving a couple of inches of twisted tape, tear off the roll and lay the twist along the cable.

Then merely make several wraps to bind the cable and the twisted tape end together.

Formerly, this was done by using two types of tape. Today, only the plastic variety is used.

When applying the tape to a joint, many people just slop it on, wrapping it in any manner.

To wrap a splice correctly, start at one end of the splice about an inch above the stripped wires and wrap to the other end of the splice, overlapping each turn by half. When you reach the opposite end, make a return wrap along the splice. Continue wrapping until you build up the tape to at least the thickness of the original insulation. To insure a neat and proper covering, keep the tape pulled taut. When you have completed the wrapping, hold the tape against the wire with your thumbnail and, with the other hand, yank the tape against the thumbnail. This will shear off the tape neatly and quickly. Then press the torn end smoothly over the wrap.

Like many electricians, you may temporarily have to stop work on a circuit before completing it. In this case, wrap the bare wire or wires (such as the cut end of a cable) for safety until you can finish the job. The trick here is to wrap the end of the cable as shown in accompanying photos. Start wrapping the tape about 5 inches from the end of the cable, holding the tape with your forefinger through the center hole of the tape. When you reach the end of the cable, turn the tape around your finger and it will twist back on itself. Merely fold this back over the end of the cable and wrap back to the start before tearing off the tape.

SOLDERLESS CONNECTORS

There are several kinds of solderless connectors, including those used for connecting heavy entrance wires, as well as small twist-on plastic connectors for connecting circuit wires to one another and to fixtures.

In using these connectors, hold the two wires parallel and close together, rather than twisting them together as described earlier in this chapter. Then you merely screw on the proper size connector to make the joint. After fastening the connector, give each wire a slight tug to make sure it is securely held in place. These connectors are very simple to install, and in case you need to do any rewiring, you merely turn them back off.

When splicing with twist-on connectors, hold the wire ends parallel. Connector threads make the bond.

Solderless connectors like this one are commonly used outdoors for heavy-duty entrance and feeder lines. To tighten, use an couple of wrenches.

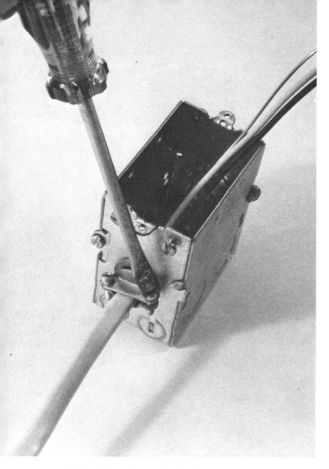

cable or conduit to the boxes. This prevents strain on the wires inside and guards against their being pulled apart, causing a short.

Plastic-sheathed cable is fastened to the boxes by various means. In some boxes the knockout hole is punched out, and a cable clamp is inserted in the hole. The cable is run through far enough to allow for the splices and connections. Then a locknut is inserted inside the box and over the cable clamp. With the cable clamp securely fast-

To hold plastic-sheathed cable in place, first mount a clamp in a knockout hole by turning down the clamp's locknut inside the box. Then run the cable through and tighten the clamp screws outside the box.

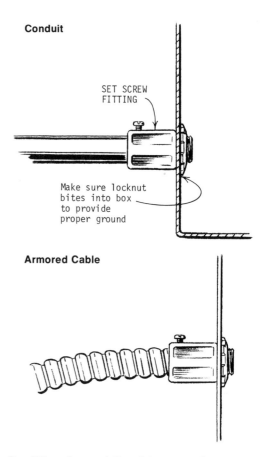

Conduit

SET SCREW FITTING

Make sure locknut bites into box to provide proper ground

Armored Cable

Box fittings for conduit and for armored cable are similar. Make sure the locknut bites into the box for proper grounding.

Solderless connectors used for heavy wires are merely clamped on the wires by means of a bolt turned down on the wires. There are other types of connectors, such as crimp connectors, but they require special equipment, and in some cases are not allowed by local codes.

CONNECTIONS INTO BOXES

The installation of the various types of boxes is discussed in chapters 9 and 10, which cover new house and old house wiring. After you install the boxes and run wiring to them, you must securely fasten the

ened to the box, you turn down the screws on either side of the clamp to fasten the cable securely in place. Some boxes have a built-in cable clamp, and in using this type of box you merely screw the clamp down in position to secure the cable.

Fastening conduit into a box is done with special fittings. Armored cable is fastened into boxes by a screw-set type of clamp, also held in place by a locknut. When securing armored cable or conduit to metal boxes, make sure the locknut bites into the box, thereby providing the proper connection for an equipment ground, if the system is to be used for that purpose.

INSTALLING RECEPTACLES

Here we'll assume that the box is in place and that the cable or wires have been run in place and clamped to the box. Next you strip away the outer sheathing and the inner paper insulation. Then strip the individual wires: Most of today's receptacles are made with push-in connectors in the back as well as the regular terminals on the sides. The push-in connectors are located to correspond with the regular "screw-type" terminals. If using the push-in connectors, strip away about ⅜ inch of insulation from each wire and push the black, "hot" wire into the hole on the side of the brass terminal. Push the white, neutral wire into the hole in the side with the silver terminal. If the circuit is to continue to another receptacle, push the other two wires into their proper holes. If for some reason you need to remove one of the wires, insert the blade of a tiny screwdriver inside the slot next to the hole and gently pull out the wire.

If you're connecting the wires to the terminals on the sides, strip about a half inch of wire and make the start of a loop. Screw out the terminal until it stops, then slip the

loop over the terminal with the loop facing the direction the terminal screw will turn. Squeeze the loop together around the screw. Then screw the terminal down tightly on the wire. Fasten a length of ground wire to the ground screw, connect it to the grounds on the other wires, and run a ground wire to a grounding screw on the metal box. These connections are described in detail in Chapter 7. Grounds are necessary on all boxes.

Push the wires back into the box and gently push the receptacle into position inside the box. Screw the receptacle to the box, making sure it is lined up "plumb," not slanted. With the receptacle in place, screw on the receptacle cover.

WIRING INTO SWITCHES

The connections on switches are made exactly the same as on the receptacles. Again some switches come with holes in the back, simplifying connecting them in. If you are using 3-way switches, make sure you know which is the *common terminal* so that the switch can be connected in properly. Switch wiring is discussed in detail in Chapter 7.

WIRING INTO JUNCTION BOXES

The Code requires that all splices be contained in some sort of junction box. After securing the box in place, run in the cables or conduit and connect them to the boxes using cable clamps or conduit connectors. Leave about 6 inches of wire protruding for connections. Strip the ends of the wire, and connect them together using either twist-on connectors or splices. Examine the wiring carefully to insure there are no open cuts or scrapes on the wires that might cause a short. Once again, run a short ground wire to a screw on the metal junction box. Push the wires down in the box and install the cover.

RECEPTACLE CONNECTIONS

Modern receptacles have holes in the back, into which you can insert wire ends for terminal connections.

Begin by attaching the looped grounding wire to the box (secured either by a screw or a grounding clip). Then run the grounding wire to the green grounding terminal on the receptacle. Insert the black wire into the hole corresponding to the brass terminal screw, the white wire into the other hole.

If you prefer to use the screw terminals, mount the wire loop so that it turns with the screw when you turn the screw down. This results in a secure connection and keeps nearly all bare wire under the screw head.

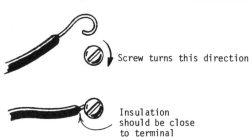

Screw turns this direction

Insulation should be close to terminal

If you wish to run the wiring to another receptacle, wire-in a second cable as you did the first—black and white wires into remaining proper holes. You can connect the second ground to ground or connect the grounds with a twist-on connector.

Before you make the final tightening turns on the screws, make sure the receptacle is vertical.

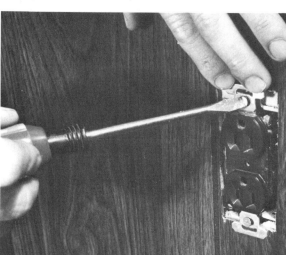

SWITCH CONNECTIONS

This is a modern 3-way switch installation. Note: There are no terminal screws, only terminal holes.

This switch offers both terminal screws and holes.

JUNCTION CONNECTIONS

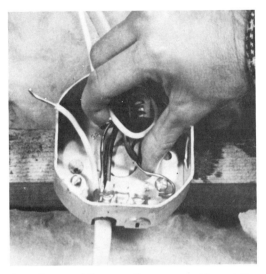

As well as providing a continuous ground through the circuit, ground wire must also be connected to the boxes. Connect to the box as shown.

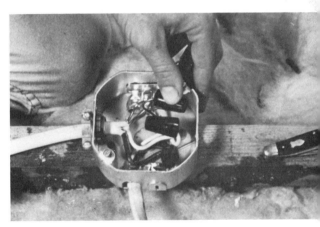

Connect the wires and then push all connections inside the box. Then fasten the box cover.

INSTALLING LIGHTS

Ceiling light boxes are available in two styles: with a center stud and without. The most common boxes come without the stud. In this case a strap, supplied with the fixture, is screwed directly to the ceiling outlet box, fastening it to the box by screws through the box cover threads.

The wires from the ceiling box are stripped and connected to the wires of the fixture, black-to-black and white-to-white. The simplest manner of connecting is to use twist-on connectors. Once the connection is made the wires are pushed up in the box and the fixture canopy is screwed to the strap, holding the fixture in place. If the ceiling box has a stud, the strap is held in place by threading a nut over the stud.

When fixtures are quite heavy, they're

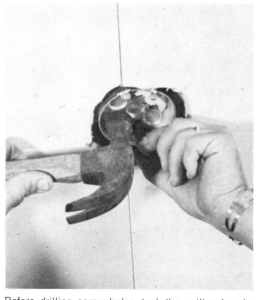

Before drilling screw holes, tack the ceiling box in place with a roofing nail, or use a screw.

CEILING-LIGHT INSTALLATION

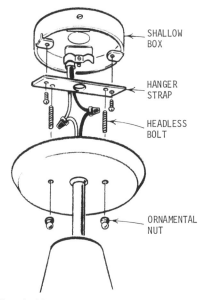

SHALLOW BOX

HANGER STRAP

HEADLESS BOLT

ORNAMENTAL NUT

You'll probably encounter parts similar to these when you install a ceiling light.

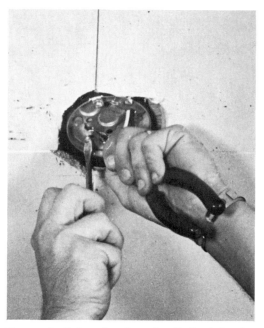

Then pull all the wires through with the help of a needle-nose pliers.

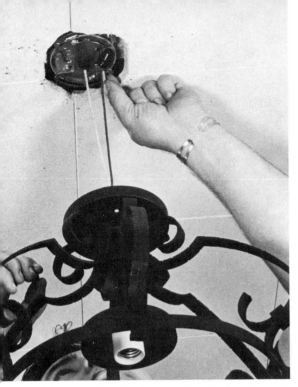

After screwing the hanger strap to the box, use twist-on connectors to secure the wires.

After securing the fixture to the hanger strap, insert the bulb and adjust the bolts for globe insertion.

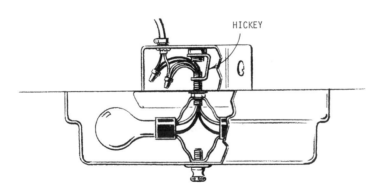

HICKEY

For heavy fixtures, you'll find a hickey-stud-nipple assembly provides reliable support.

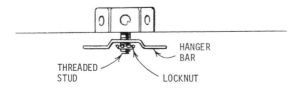

HANGER BAR

THREADED STUD

LOCKNUT

Some stud and hanger-strap assemblies employ a locknut.

often installed to the stud in the ceiling box by using a hickey. The initial treatment of the installation of a chandelier, as well as a chain-hung fixture, is the same except for the mounting of hardware.

Wall fixtures are installed in the same manner as ceiling fixtures. You can use either the stud or hanger bar to mount the fixture to the box.

There are two types of fluorescent fixtures: the starter type and the ballast type. Both are normally fastened to an outlet box by means of a stud.

One of the more popular new types of fixtures is the recessed light. This may have "spot" lights which are recessed in the ceiling or a series of fluorescent tubes mounted inside a reflector and covered by a plastic diffusion screen. The spot lights normally come prewired in a shell including a junction box into which the circuit cable is run. The fixtures are mounted to the studs, using factory installed, sliding adjustable bars which enable you to easily mount the fixture. The knockout hole is removed from the box on the light, the wires connected in, the bulb screwed in place, the glass plate installed, the cover mounted.

FLUORESCENT-LIGHT INSTALLATION

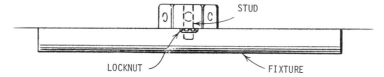

The stud and locknut assembly is often used for fluoresent fixtures.

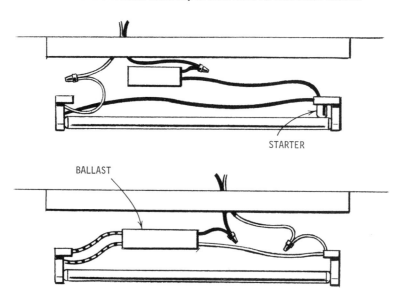

If you install fluorescent lights, you'll probably encounter either of these two types of wiring schemes.

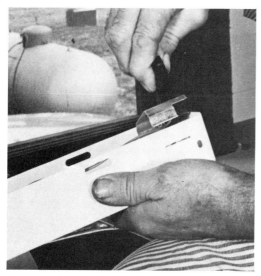

After snap-fitting the box together, you can insert the tube holders, as shown.

After positioning the fixture, pull the wires through and screw the unit into place.

Mount the bottom side of the box, as shown. And then insert tubes — and a shade, if desired.

7

Circuits

There are literally thousands of combinations of circuits, each suited to a specific situation. But using as guides the basic circuits described in this chapter, you'll be able to assemble the combinations needed for your own wiring jobs. If you have made a good plan of the circuits needed, as explained in Chapter 3, you already know what types of circuits you want.

Some local codes may require that all wiring in new work (new homes) be run through thin-wall conduit, requiring that you pull the wires through the conduit to make up the circuit. This actually is the easiest method, when it comes to simplifying and understanding a circuit, because you can pull through the exact number of proper colored wires needed. (Some circuits are more easily installed using four wires.)

Plastic-sheathed cable comes in combinations of only 2 or 3 wires: black and white; black, white and red; or black, white and green (used as a ground); or black and white combined with a bare ground wire. On some circuits using plastic cable, you may have to break the one cardinal rule

of wiring: *Always run the white or neutral wire direct from the source, without interruption, to the point where the current is to be consumed at 115 volts* (such as a light fixture). On certain circuits using plastic-sheathed cable you may break this rule. But in this case, it is wise to paint or tape the white wire black at any location where it is connected to a black wire, whether at a switch, light or other fixture. This is a precaution against accidental shock years later.

The connections wired with 2- or 3-wire plastic cable can be somewhat confusing because of this. So all drawings in this chapter show the more difficult to install plastic cable. Then it's a simple matter to use the same drawings as guides for installing circuits of thin-wall conduit, just pulling the right colored wires through.

In all circuits shown, the boxes for the receptacles and switches have already been installed properly, as shown in Chapter 6. All that needs to be done is to run the wires and hook up the circuits. All connections are illustrated with twist-on connectors for the wire splices.

Most codes require that all wiring, lights, switches, boxes, and receptacles be grounded. The first illustration here shows the proper method of installing a junction box and connecting ground wires, *as well as connecting a ground jumper wire directly to the box using a bonding screw in one of the threaded screw holes.* With this method, should lights or receptacles be removed from the boxes, the ground continuity would remain, as specified by the Code.

For the sake of clarity, only the initial junction box illustration and the several illustrations on receptacles show grounds. With the ground wires omitted from the illustrations, you'll find it easier to understand the connection of the various colored wires. Just remember to connect the ground to ground, and to each box it passes through.

Installing a Junction Box. In most circuits you will probably have to install junction boxes to connect more than one outgoing source line to the incoming source. The illustration shows the proper connections for wiring up a junction box with a source and power going out to two separate locations. The Code and most local codes restrict the use of the boxes to not more than a specified number of wires according to box and wire size. Refer back to Chapter 5 for more on box sizes.

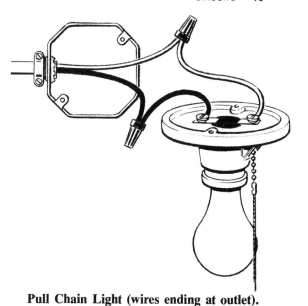

Pull Chain Light (wires ending at outlet). This is the simplest circuit to wire, and is normally used in closets, pantries, basements and attics. These fixtures may either have two screws for connecting the wires to the fixture or two wires leading from the fixture. If the fixture is equipped with screws only, bring the black wire direct from the source into the ceiling box and connect it to the brass screw. Bring the white wire directly from the source and connect it to the silver screw. If the fixture has two wire leads instead of the screw terminals, connect the black to black and the white to white wires

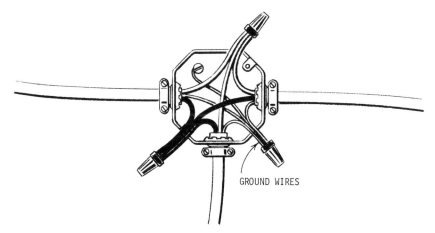

GROUND WIRES

according to local codes. This may require either twisting together and taping, soldering and taping, or using twist-on threaded connectors or insulated crimp connectors. Then connect the grounds, not shown, and install the fixture.

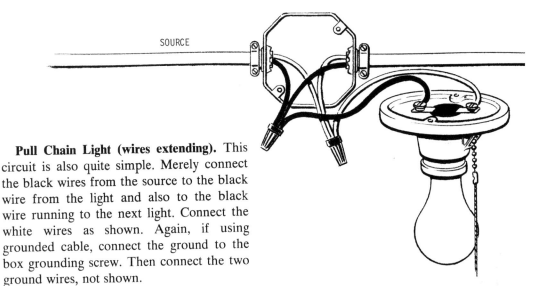

SOURCE

Pull Chain Light (wires extending). This circuit is also quite simple. Merely connect the black wires from the source to the black wire from the light and also to the black wire running to the next light. Connect the white wires as shown. Again, if using grounded cable, connect the ground to the box grounding screw. Then connect the two ground wires, not shown.

Ceiling Light Controlled by Wall Switch (thin-wall conduit). In this instance we show how simple it is to keep the colors of the wires properly connected by running thin-wall conduit. Note the two black wires running from source and outlet to switch. Naturally this wouldn't be possible with plastic cable, shown at the bottom of the next page.

The black wire from the source is connected to a black wire run from one terminal of the switch. A second black wire from the switch is run to the black wire or brass terminal on the light. The white wire from the source is connected to the white wire of the light, or silver colored terminal, thereby making a continuous run of the white wire from the source to the outlet. Connect grounds, not shown.

Ceiling Light Controlled by Wall Switch (using plastic cable). This illustration shows the same essential hardware that you saw in the one above it, yet the wiring is different. The Code allows the use of white wire as the hot wire only in this instance. Note that the white wire is painted black at the connection to indicate it is actually a "black," or hot wire. Connect the black wire from the source to the white wire running to one side of the switch. Run the black wire from the switch to the black wire or brass terminal of the light. Connect the white wire from the source to the white wire of the light, or to the silver terminal. Connect grounds, not shown.

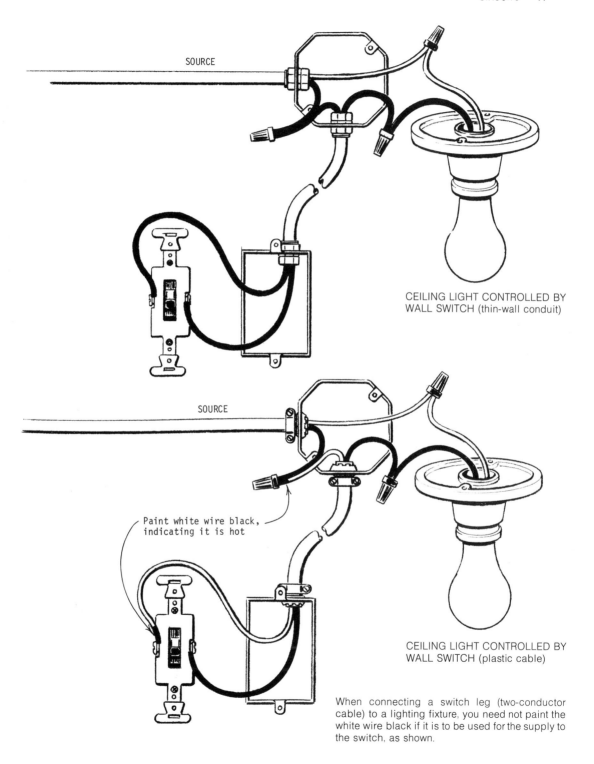

SOURCE

CEILING LIGHT CONTROLLED BY
WALL SWITCH (thin-wall conduit)

SOURCE

Paint white wire black,
indicating it is hot

CEILING LIGHT CONTROLLED BY
WALL SWITCH (plastic cable)

When connecting a switch leg (two-conductor
cable) to a lighting fixture, you need not paint the
white wire black if it is to be used for the supply to
the switch, as shown.

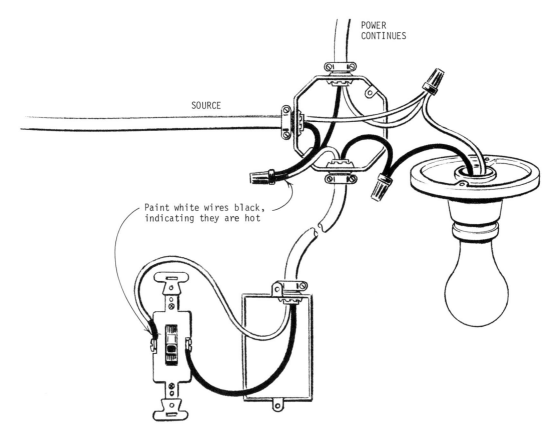

POWER
CONTINUES

SOURCE

Paint white wires black,
indicating they are hot

Ceiling Light Controlled by Switch (power extending past light). There is little difference between this circuit and the two just preceding except that here the power extends past the light, perhaps to a light with a pull chain switch on it, or even to a receptacle. If the power extends to a receptacle, make sure you use a 3-wire grounded wire cable, grounding to the boxes and to the receptacle. The grounded wire is not shown in the illustration.

Once again, in using nonmetallic plastic-sheathed cable, you must paint the white wires (running from the switch to source) black on each end to indicate a hot wire. This will help prevent accidents in future switch repairs or replacements. Note that in two cases you'll have 3 wires connected together. If using twist-on connectors, make sure you use a large enough connector to insure the wires are tightly and securely fastened together. Also check your local code for the size of box required when this much wire must be installed. Note that the white source wire runs directly to the white wire running past to the next light. The black wire connects to the black wire running past and to the white wire painted black, which runs to one side of the switch. The black wire from the switch is connected to black wire of the light receptacle. (Running this circuit through conduit, it would be a simple matter to pull through a black wire instead of the white and to wire all black together and all white together.) Connect grounds, not shown.

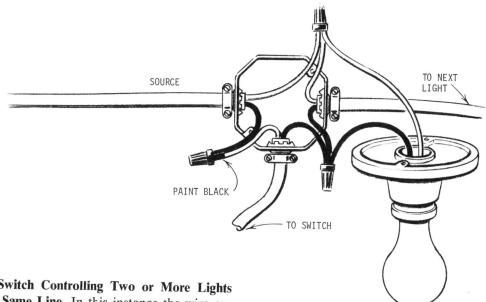

SOURCE

TO NEXT LIGHT

PAINT BLACK

TO SWITCH

Switch Controlling Two or More Lights on Same Line. In this instance the wire extending past the light runs to another light controlled by the same switch. Turning on or off the switch controls both lights, or any number of lights. For this very simple circuit, you merely wire the first light to the switch and then run power to the next light. Once again all white wires are connected, and the black wire running from the switch is connected to the black outgoing wire and the black wire from the light terminal. When connected to the black source wire and run to the switch, the white wire is once again painted black at connectors to indicate a hot wire. Make sure you paint it black at both ends. Connect grounds, not shown.

Switch Controlling Light (power through switch). This circuit (shown on page 80) differs from the previous ones in that the source comes through the switch. In some instances it may be easier and more economical to run the source wire to the switch first, rather than to the box location of the light. In this circuit the white wire is connected white-to-white in the switch box. The black

wire from the source is broken to make the connections on the switch. The light is wired white-to-white and black-to-black. Connect grounds, not shown.

3-WAY SWITCHES

Wiring 3-way switches is a confusing wiring task for the novice, and it sometimes can be confusing to the experienced electrician as well. This is because there can be many different combinations of switches and lights. For instance, the source could run to switch, then to outlet, then to another switch. Another combination might be from source to switch, then to a second switch, then to outlet. A third combination might be from source to outlet, then to switch and finally to a second switch.

As you can see an infinite number of circuits can be run with 3-way switches, depending on the best location of the source and on locations of switches and lights. You

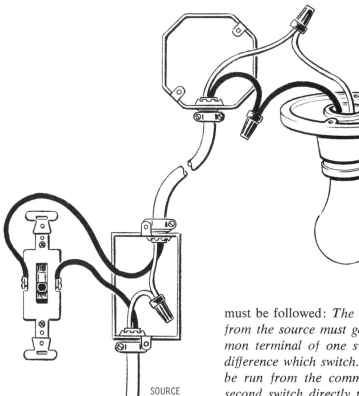

SOURCE

must use switches that are especially made for use in 3-way circuits. These switches will normally have three terminals. One will be a common terminal, which may be marked by a different color than the others. Or in case of push-in switches, the common terminal may be marked by an arrow or by some other means. Make sure you know which terminal is the common one, because that is the most important element in wiring 3-way switches correctly. In all the drawings shown, the common terminal is on the side of the switch opposite the other two terminals, but this may not be the case with some switches. Once again, make sure you know which is the common terminal.

In wiring all 3-way switches, one rule must be followed: *The incoming black wire from the source must go direct to the common terminal of one switch. (It makes no difference which switch.) A black wire must be run from the common terminal of the second switch directly to the black wire or brass terminal of the light fixture.* If you remember this one rule, you'll be able to wire 3-way switches without any problems. It's a good idea to lay out the position of your switches and lights on paper and mark the circuit using colored pencils.

You will have to use a 3-wire cable for wiring 3-way switches. In most cases you can use a 3-wire cable with black, white and red, *not green*. After running the connecting black wires, connect the white incoming wires direct to the white wire or silver terminal of the light fixture, and run the red and black wires of the 3-wire cable to connect the remaining two terminals of one switch to the remaining two terminals of the second switch. Just remember that the Code specifies that *the wires at the fixture must be black-to-black and white-to-white.*

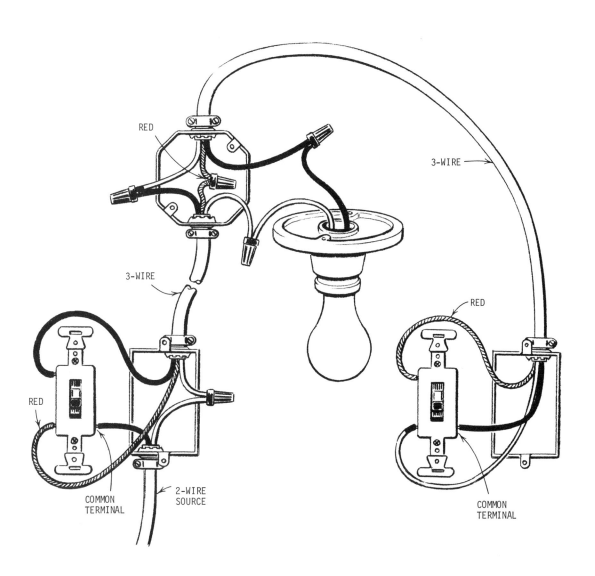

RED

3-WIRE

3-WIRE

RED

RED

COMMON
TERMINAL

2-WIRE
SOURCE

COMMON
TERMINAL

3-Way Switch, Controlling Light in Middle of Run (power through switch). As is typical, the black wire from the source, running direct to the first switch, connects to the common terminal. The white wire runs directly from the source to the white wire of the fixture. The black wire from the common terminal of the second switch runs directly to the black wire of the fixture, and the remaining wires are connected to the remaining terminals as shown in the illustration. Connect grounds, not shown.

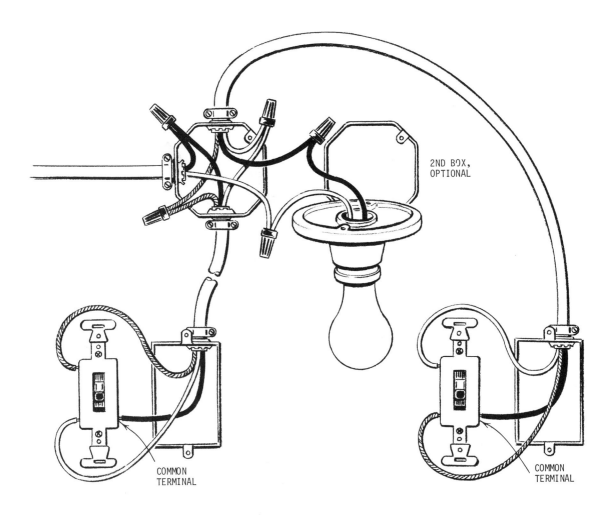

2ND BOX,
OPTIONAL

COMMON
TERMINAL

COMMON
TERMINAL

3-Way Switch, Controlling Light in Middle of Run (power from light box). Although it appears more complicated, this 3-way circuit is much the same as the previous one, except for the number of splices in the junction box. Because of the number of connections, you may need to install two boxes, one for the light and, if possible, one a short distance from the light for the connection of the switches. Here you run both the black and the white wires from the switches, through their boxes to the fixture. To simplify illustration, ground wires are not shown here, but they are recommended.

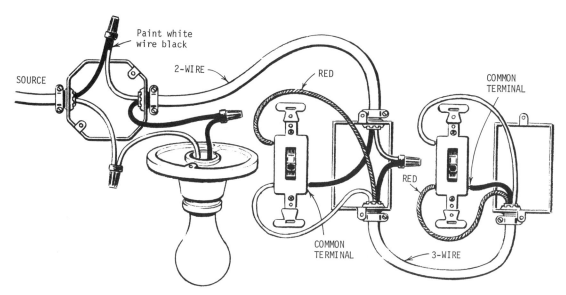

Paint white
wire black

SOURCE

2-WIRE

RED

COMMON
TERMINAL

COMMON
TERMINAL

RED

3-WIRE

3-Way Switch, Controlling Light Ahead of Switches. In this circuit you will once again have to resort to using a white wire as a "feed-through wire" if using nonmetallic, plastic-sheathed cable. This is permitted by most codes if the white wire is painted black, both at the switch and at the box where it connects to the black source-wire. To simplify illustration, ground wires are not shown here, but they are recommended.

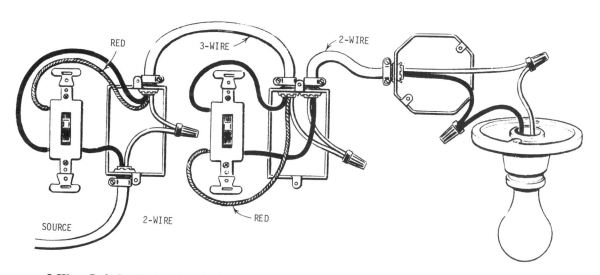

RED

3-WIRE

2-WIRE

SOURCE

2-WIRE

RED

3-Way Switch Controlling Light on End of Run (power through first switch). This is also a very basic switch circuit. Again, you connect the black from source to the common terminal, connect all white wires, and connect the black wires from the common terminal of the second switch on the way to the fixture. Connect grounds, not shown.

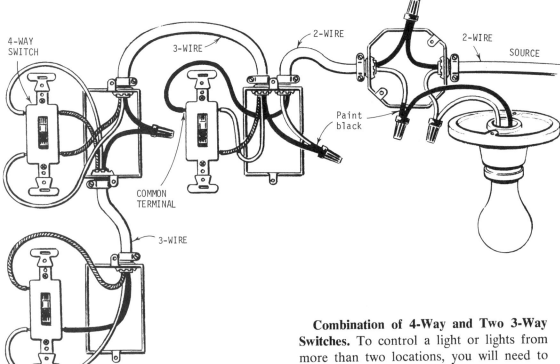

4-WAY
SWITCH

3-WIRE

2-WIRE

2-WIRE

SOURCE

Paint
black

COMMON
TERMINAL

3-WIRE

Combination of 4-Way and Two 3-Way Switches. To control a light or lights from more than two locations, you will need to install one 4-way switch as well as the 3-way switches. This is somewhat complicated if you use nonmetallic plastic-sheathed cable. Just remember that you should follow the basic rule of connecting the common terminals, except that in the last connection from the common terminal there will be a white wire running to the black wire of the fixture. Again paint each end of the white wire black to indicate that the wire is a feed-through, or "hot," wire.

You'll need a 4-wire system in order to connect the 4-way switch in the circuit. And because 4-wire cable is not manufactured, you'll find it much easier to install these circuits using conduit. This allows you to pull the correct colored wires to the switch locations. However, you can install a 4-way with cable, using the drawing shown, provided your local code allows. Connect grounds, not shown.

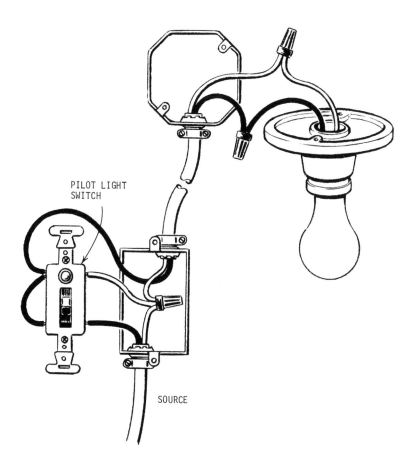

PILOT LIGHT
SWITCH

SOURCE

Pilot Light Switch Controlling Light.
Many newer switches contain a small pilot
light or a light in the switch handle itself that
shows you where the switch is in the dark.
Wiring some of these switches is a bit different from wiring a standard switch. Make
sure you understand the arrangement of the
terminals on the brand of switch you use.
You'll find the wiring diagram on the box
containing the switch. The illustration shown
here gives only one type of wiring diagram

for one type of switch, and does not show
grounds. Check to be sure exactly how your
lighted switch should be installed.

When using plastic-sheathed cable, you'll
find it easier to install the switch with the
source coming directly through the switch
rather than from the light and then to the
switch. With conduit, the arrangement can
be in any order because you can pull the
correctly colored wires through to make the
circuit. Connect grounds, not shown.

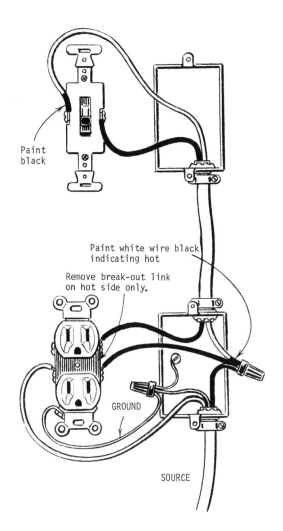

Paint black

Paint white wire black indicating hot

Remove break-out link on hot side only.

GROUND

SOURCE

To modify a receptacle to 2-circuit, break off the link connecting the hot terminals.

It is advisable to wire these circuits by using 2-circuit receptacles. Most better receptacles are made so they can easily be modified to 2-circuit by breaking off the connecting links connecting the hot terminals, as shown in the photo. In these receptacles, one half of the unit is controlled by a switch while the other half remains constantly hot. It's a good idea to install these receptacles with all the switch halves at the bottom or at the top, so you'll always know which is constantly hot.

Wiring this type of circuit with plastic cable is not particularly hard, as shown in the illustration. Again, paint the white wire black at both the receptacle location and at the switch location. This circuit is also easiest installed with the source coming through the first receptacle, as shown.

Switched Receptacles. In some rooms, you may wish to have some receptacles controlled by a switch so that you can turn on lamps when you enter the room. These can be controlled by a 3-way switch at either end of the room. Other receptacles for appliances such as clocks, radios and television can be left unswitched.

Single Receptacle Outlet, Grounded. All receptacle outlets should be grounded, and only grounded receptacles should be used. Any old ungrounded receptacles should be replaced with new grounded wire and grounded receptacles where possible. Or a ground should be run from the receptacle, as described in Chapter 8.

Connect the black wire from the source to the brass terminal screw, and the white

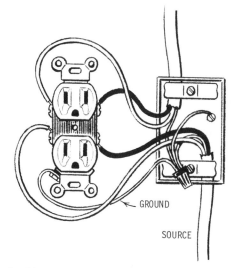

GROUND

SOURCE

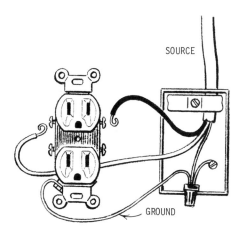

SOURCE

GROUND

wire from the source to the silver terminal. Connect the ground wire, which may either be a bare wire or green, to the hexagonally shaped green terminal or terminal otherwise marked as the ground. It's a good idea to also run a ground wire to the box itself. If you have to remove the receptacle later, the box will provide a ground continuity as required by the Code.

Receptacle Outlet, Grounded (wires running through to another receptacle). Any number of receptacles can be wired in this manner, depending of course on the amount of load and the number of receptacles allowed by code rules.

Receptacles for use in this circuit are re-

quired by the Code to contain two brass and two silver terminals. Run the black wire from the source to the brass terminal. Connect the outgoing black wire to the other brass terminal. Connect the white wires in the same manner. Connect both grounding wires to the outlet box, and run a short grounding loop to the green terminal. Connect the additional receptacles in the same manner.

It is possible to make a short loop in a wire, strip the loop and fit it under the terminal (making one continuous wire through the receptacles).

DIMMER SWITCHES

One of the most enjoyable as well as practical electrical devices is a "dimmer," or brightness control switch. By utilizing these units on an overhead chandelier in a dining room, or even for the main light of a living room, you can give a room a soft romantic glow or various brightnesses—all by merely turning a knob activating a tiny rheostat in the switch.

Dimmer switches are made for both incandescent and dimming-type (special ballast) fluorescent lighting, and in many different wattages. So make sure you use the proper switch for your situation. A small amount of heat is normally generated in these switches; so if you install two switches in a double box, make sure you limit the wattage. For instance two 600-watt switches should be limited to a maximum load of 500 watts. If using 3 switches in the same box, limit the wattage to 300. Dimmer switches should be used only for permanently installed light fixtures. For motors, you can buy specially designed rheostat controls. Purchase only quality, brand-name switches. Otherwise the switch may cause radio or television interference.

Installation of a dimmer switch is the same simple operation as the installation of a standard switch, regardless of whether the switch is a single-pole or a 3-way switch. The most important point to remember when installing the switch is to shut off house power or power to the applicable circuit. This is

for two reasons: First, safety-wise, it is foolish to work on any hot circuit. And in addition, the electronic components of a dimmer switch may easily be damaged during installation if the circuit is "alive."

After turning off the power, remove the wall plate and the old switch. Attach the two source wires in the wall box to the terminal screws of the dimmer. If the source wires also include a ground wire, connect the ground to the metal wall box by use of a grounding screw. Mount the dimmer unit in the box and turn down the holding screws to fasten the unit in place. If the box is grounded, this automatically grounds the dimmer switch unit. Pull the control knob off the switch, and replace the wall plate with any normal switch plate that matches the color of the control knob. Merely push the control knob back onto its spindle and turn the power back on.

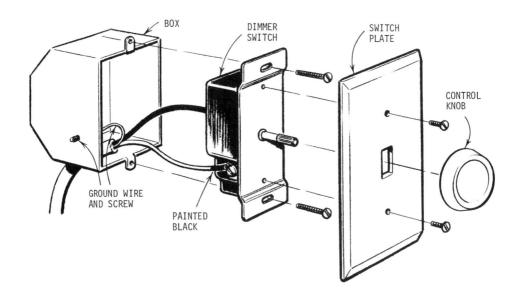

Service Entrance and Grounding

The wires coming in from the utility company's transformer or lines are called *service drop* when used in an overhead service and *lateral* when used in an underground system. They are connected to the house electrical service equipment (meter and main switch or panel) by means of service-entrance wires. The power company wires include 1) a black insulated wire, 2) a red insulated wire (or another black), and 3) a neutral ground wire, which may be a white insulated wire but usually consists of twisted strands of bare wire. Years ago these wires were strung individually from the utility pole to the house, but today a twisted, combined-wire system is considered safer because it provides more strength over long spans. The combined wire also poses less danger from "shorts" when the wind whips the wires about. In theory, the black wire and the red wire are hot, but you should consider the neutral (ground) wire hot also.

Once you bring these three wires into a service panel, if you connect the black wire and the neutral in a circuit, you will have 120-volt service. The same holds true when you connect the red wire and the neutral. But if you connect the black wire, the red wire, and the white neutral, you will have 240-volt service, which you may need for heavy-duty appliances such as water heaters, clothes dryers, ranges, and some air conditioners.

There are six major components in a service entrance:

1. The service wire from the utility company to the building or pole.
2. The service-entrance cable, raceway (conduit) or mast, with individual service-entrance wires.
3. The meter socket.
4. A complete disconnect, which allows you to disconnect all power coming in from the utility wires.
5. Overcurrent protection in the form of fuses or circuit breakers.
6. The ground.

There are several mechanical means of installing a service entrance, and they are selected according to the house or pole location, as well as according to local codes. This is one aspect you should very definitely check with local utility companies because they're very strict as to what type of service entrance installation is allowed. At one time, meters were installed indoors, but today most companies require that the meter

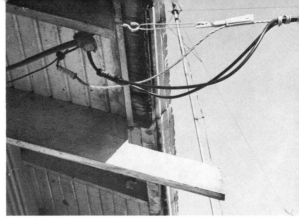

Most modern service entrance installations made by utility companies have an "insulator connector," as shown, that brings in a single line of 3 twisted wires. The bare, supporting wire is the neutral.

This old, 3-wire installation presents 2 definite hazards. First, 2 of the drip loops don't droop enough to prevent water from running into the service head. Second, the head's mounting screw is loose.

be installed outdoors and in an easily readable location.

The four main types of installation are 1) conduit, 2) mast, 3) service entrance cable, and 4) underground cable. Often the selection of the type of installation will be governed by the location of the service drop wires from the power company. The Code states that service drop wires must be located above ground as follows:

1. At least 10 feet above finished grade, sidewalks, or projections from which they might be reached.
2. At least 12 feet above residential driveways and commercial areas not subject to truck traffic.
3. At least 15 feet above commercial areas, parking lots, truck traffic areas, and agricultural areas.
4. At least 18 feet above all public thoroughfares such as streets and alleyways.

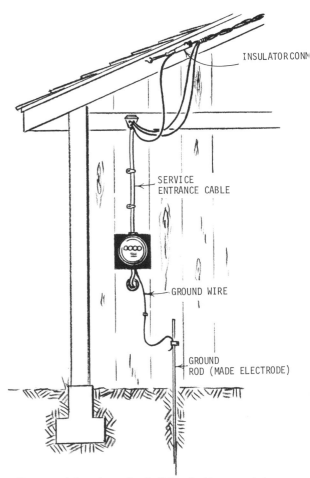

To ground the wires of a building lacking a metal water pipe system, run a copper wire to a ground rod (made electrode).

3 WAYS UTILITY COMPANIES SECURE ENTRANCE WIRES

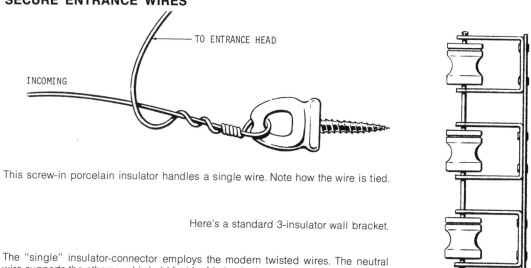

TO ENTRANCE HEAD

INCOMING

This screw-in porcelain insulator handles a single wire. Note how the wire is tied.

Here's a standard 3-insulator wall bracket.

The "single" insulator-connector employs the modern twisted wires. The neutral wire supports the others and is held fast by friction in the connector. The drip loops prevent rain water from running into the entrance head.

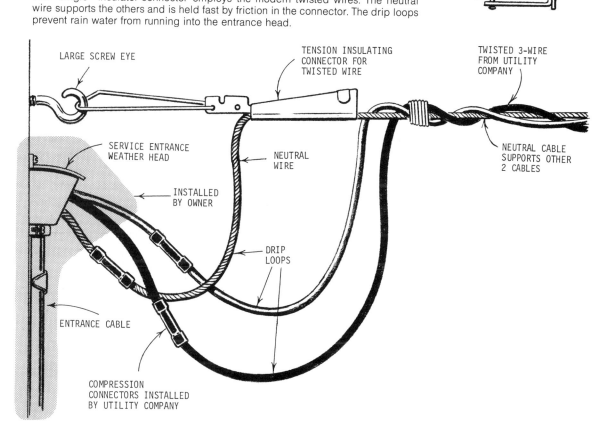

LARGE SCREW EYE

TENSION INSULATING CONNECTOR FOR TWISTED WIRE

TWISTED 3-WIRE FROM UTILITY COMPANY

SERVICE ENTRANCE WEATHER HEAD

NEUTRAL WIRE

INSTALLED BY OWNER

NEUTRAL CABLE SUPPORTS OTHER 2 CABLES

DRIP LOOPS

ENTRANCE CABLE

COMPRESSION CONNECTORS INSTALLED BY UTILITY COMPANY

In addition, the wires must be at least 36 inches from all windows, doors, porches, fire escapes, and other similar places. The Code considers conductors run above the top level of a window out of reach from that window and may be closer than 36 inches. When the wires pass over a roof, they must be at least 8 feet above the highest point on the roof. They may be 3 feet where the voltage is less than 300 and the roof has a slope of not less than 4 inches to the foot. Where the voltage is less than 300 and the service-entrance wires are installed in conduit (a mast), the wires may pass within 18 inches of the roof, provided they do not pass over more than 4 feet of roof overhang.

The insulators should be installed separately. If you place the insulators above the entrance head, make sure you leave enough wire for drip loops that allow rain and melt water to run off the wires instead of into the head.

In the following pages, we will discuss the installation of the four major types of service entrances, then the installation of a fuse box, then a circuit breaker box, and finally grounding installations for both urban and rural situations.

INSTALLATION OF SERVICE ENTRANCE CABLE

This is the easiest type of service entrance to install. However, local codes may disallow it. So make sure you check beforehand. The accompanying drawing shows a complete entrance cable installation, indicating the wiring procedures through the meter and into the service panel located in the house.

The first step is to cut away the outer sheathing from the service cable. Then separate the bare neutral wires and twist them together to make one single twisted wire. Disassemble the entrance head and insert the three wires through the opening in the head, leaving about 36 inches of wire protruding for the connection to the service wires, which will be connected-in when the service entrance is finished. Screw the service cap back together.

Place the entrance head in the proper position, as high as possible, and according to code rules, and fasten it in place by driving screws through the screw holes provided.

Next, determine the location of the meter housing and bore a hole through the wall below the meter to run the lower portion of the entrance cable into the house. Fasten a watertight connector, specially made for service entrance cable, at the top of the meter base and insert the cable. Leave about 10 inches of cable protruding through the housing for connecting into the meter. (Some electricians like to leave the bare neutral wire as a single wire running completely through the meter, *but connected,* and into the entrance panel. But this makes for a great deal more work and care in installing the housing.)

Connect another length of cable on the bottom of the meter by using another watertight fitting and run it into the hole bored through the wall. Again leave plenty of cable for connecting to the service panel on the inside of the building. Install a sill plate fitting where the cable enters the building to keep out water. Another method of preventing water from running into the service panel is to use a drip loop as shown earlier in this chapter. Fasten the meter housing in place by screwing it to the wall.

Strip the ends of the wires coming down from the top and connect to the top terminals. Strip the ends of the wires coming in from the bottom of the housing and again fasten them on the terminals. It takes quite a bit of pressure to turn the terminals down on wires of this size, so use a large screwdriver. Connect the twisted bare neutral wires from both top and bottom to the center screw. Using the side of a hammer, or the handle end of a screwdriver, gently push the wires into the housing, making sure there are no places where bare wires might cause shorts.

The installation is now ready for a ground which may be connected to either the neutral terminal of the meter, if you are using an outside ground, or to the neutral terminal bar of the service panel, if you want to run the ground inside. The entrance wires can now be connected to the service panel, as explained later in this chapter.

WHEN SERVICE ENTRANCE CABLE IS ALLOWED

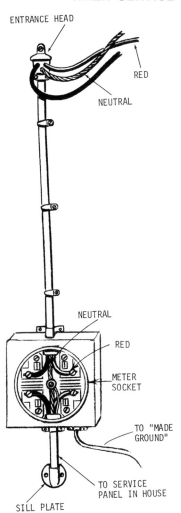

ENTRANCE HEAD

RED

NEUTRAL

NEUTRAL

RED

METER SOCKET

TO "MADE GROUND"

TO SERVICE PANEL IN HOUSE

SILL PLATE

From the entrance head, bring the wires down to appropriate meter terminals. From the meter, all wires continue inside to the service panel.

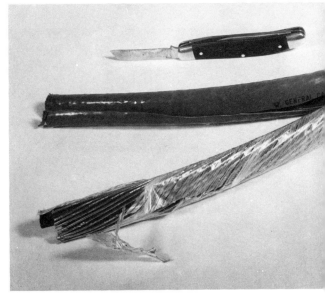

To begin, strip service entrance cable with a knife, removing the outer and the inner coverings.

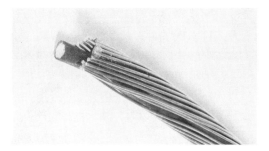

If you look close here, you can see the ends of both the red and the black wires. (The red is partially hidden by the strands of bare neutral wire.)

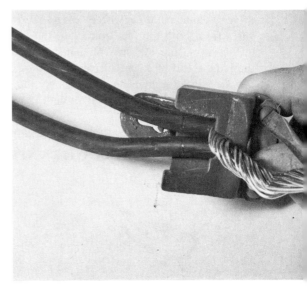

Twist the neutral strands to make up the neutral wire, as shown left. (Note the entrance head components.) Then bring the 3 wires up through the bottom of the head, leaving the edge of the sheathed portion flush with the bottom inside of the head's cable clamp. Secure the clamp, and bend the neutral wire into the bottom slot.

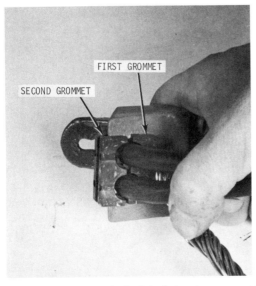

FIRST GROMMET

SECOND GROMMET

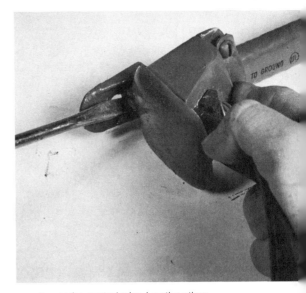

As shown in the left photo, place one rubber grommet over the neutral wire. Lay the other 2 wires into the 2 niches in the first grommet, and then place the second grommet over the 2 wires. Then, with the wires in place, screw down the head cover, while holding the wires and grommets in position.

Mount the entrance head and cable according to local code. Make sure you have enough cable to reach the meter base.

Fasten the meter head to the building. Local code may allow a drip loop like the one shown here.

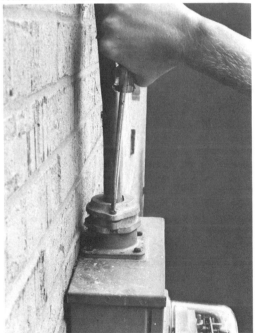

Secure cable through the top and bottom of the meter housing by means of watertight connectors. Be sure there's enough cable to connect to the panel inside.

With the meter secured, strip the wires and fasten them to the proper terminals. Fasten the neutral wire to the center terminal.

Fasten the remaining wires as shown here.

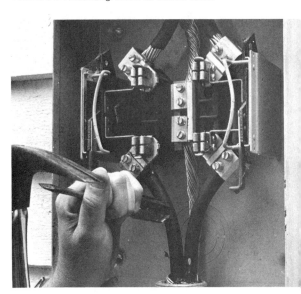

Force the heavy wires neatly into the meter by driving the smooth handle of a screwdriver against them.

INSTALLING A
CONDUIT SERVICE ENTRANCE

A conduit service entrance is installed in much the same manner as a cable type is, except that conduit piping and special fittings are used and the wire is pulled in place after the fittings are installed. The accompanying drawing shows a typical installation of a conduit service. After installing the service insulator or insulators, you should mark the approximate location of the meter, and bore the hole in the building for the service entrance ell (conduit body) to enter. Measure the distance from the location of the meter to the hole.

The assembly consists of a service head, a section of conduit, the meter housing, another section of conduit and a service entrance ell to bring the wiring into the house. Temporarily position the meter and the service ell in their places. Measure between the two and cut a piece of conduit to fit between them. Measure from the top of the meter to a point at least a foot above the incoming wire insulators. Ream the conduit to remove any burrs, and assemble the parts on the ground. Install the assembly on the side of the house, pushing the short end of the ell into the hole in the side of the house. Install a short piece of conduit to the entrance hole on top of the service panel and install a locknut and a grounding bushing. Fasten the entire assembly to the wall using pipe straps as shown.

With the mechanical assembly in place, you're ready to pull the wires through. Pull through the proper size wires for the installation according to the charts shown in Chapter 2. Pull through a red wire, a black wire (or two black wires), and a white wire for the neutral wire. Cut enough wire to reach from the meter socket to the service

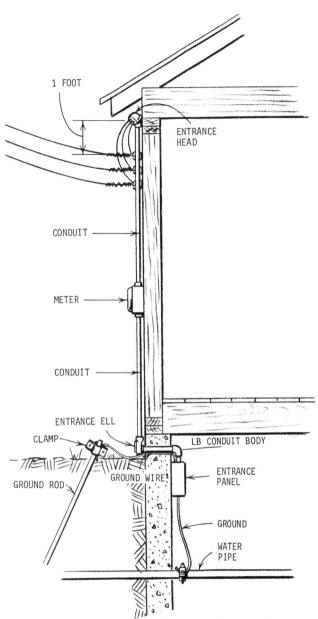

This is a typical conduit service installation. If at all possible, the entrance head should be 1 foot above the topmost service wire. (If the water supply pipe is plastic, the pipe electrodes must be supplemented by an additional electrode.)

head, plus about 36 inches extra, and then remove the top of the service head. Pull the wires into place. Then thread them through the holes and install the service head top. Make sure you cut the wires long enough to reach from the meter socket into the service panel. The Code doesn't allow any splices, but clamped or bolted connections in metering equipment enclosures are permitted. By removing the cap on the entrance ell you can thread the wires through the elbow much more easily.

INSTALLING ENTRANCE CONDUIT

1. On entrance conduit, use a hacksaw to cut the lengths you'll need above and below the meter.

2. Ream the conduit's inside edges to remove burrs that could cut into insulation on wires.

3. Mount the entrance head and install the "stack" to the meter. Threading the wires comes later.

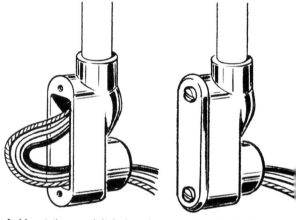

4. Mount the conduit below the meter and slip on the entrance ell. As in this illustration, remove the ell's cap to help bring the wires through.

INSULATED-THROAT
GROUND BUSHING
WITH GROUND LUGS

You'll need a grounding bushing instead of a regular
locknut connector in order to connect the conduit to
the service panel inside.

The entire conduit assembly is held in place by conduit straps, such as these.

MAST INSTALLATION

A mast installation is essentially a conduit
service, except that it is extended through
the roof of the house. Such masts are often
used on low-roofed houses when it's not possible to achieve the needed height for the
incoming service wires. A service mast is
normally sold in kit form with all the needed
materials including roof flashing to make a
seal around the conduit. You will also have
to purchase a meter hub so that you can install the conduit in the service panel. After

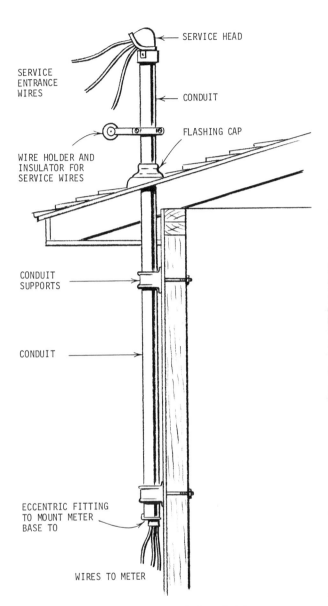

SERVICE HEAD

SERVICE
ENTRANCE
WIRES

CONDUIT

FLASHING CAP

WIRE HOLDER AND
INSULATOR FOR
SERVICE WIRES

CONDUIT
SUPPORTS

CONDUIT

ECCENTRIC FITTING
TO MOUNT METER
BASE TO

WIRES TO METER

This is a typical mast-type installation—that is, a
conduit assembly that goes through the roof.

installing the meter and the lower conduit, just as when installing conduit service, it's merely a matter of pulling the wires into place and connecting them in.

The use of a service mast allows a low clearance of the incoming wires. According to the Code, if the mast is within 4 feet of the roof overhang, the wires need clear the roof only by 18 inches from the nearest point.

UNDERGROUND SERVICE ENTRANCE

Today a great number of modern homes are rigged with underground service. The start of an underground cable system is very similar to a regular conduit system. It begins at the service pole with a conduit run to the top of the pole, fitted with a service head. The conduit is run down the side of the pole and at least 24 inches below grade to protect the wire or cable. A special bushing is fitted into the end of the conduit, and the wiring is threaded through the conduit, out the bushing, and along the trench to the house. The burial depth for the cable would normally be 24 inches deep. At the house, install the conduit down into the trench. Then bring the wire up through the conduit and into the meter housing.

The only type of wire that can be used for this underground entrance is type U.S.E. (underground service entrance). It is very tough and water resistant.

CONNECTING TO THE SERVICE PANEL

Regardless of whether the service panel is a fuse panel or a circuit breaker, the installation of the service entrance wires is basically the same as shown in illustrations on upcoming pages. To prevent getting a severe shock, *never wire into a panel with the power service lines connected in. Always wire the panel; then have the utility company connect their incoming wires to the wires protruding from the service head.* From that point on, all work can be done on the panel by merely pulling the main fuses or turning off the main circuit breaker.

A service panel serves two purposes. First, it is a means of manually shutting off all electrical power or particular circuits for working on them. Second, it acts as an overcurrent protection so that a short or circuit overload will blow a fuse or trip a circuit breaker, causing the current to be "killed." Always use the proper size service panel for your particular installation. Chapter 2 describes how to determine this. Service panels should be marked "suitable for use as service equipment."

The service panel should be located securely and in a good easy-to-get-at place. It should be at least three feet from the ceiling and preferably next to a door, such as a garage door. Check with local authorities as to the proper location in your area.

To connect the three wires into the service panel, place a nonwatertight connector over the cable. (If you are using conduit, you'll need a special conduit fitting designed for service entrance panels.) Then pull the wires into place. Strip the two incoming wires and tightly twist the bare strands of the third wire together. Next, connect the neutral bare wire of the service directly to the neutral busbar in the metal cabinet. Connect the two hot wires to the main lugs on the main disconnect. (These are normally marked mains.) The busbar will also have a connector for the grounding electrode wire, as well as many terminals for the neutrals and grounding wires of all the 115-volt circuits.

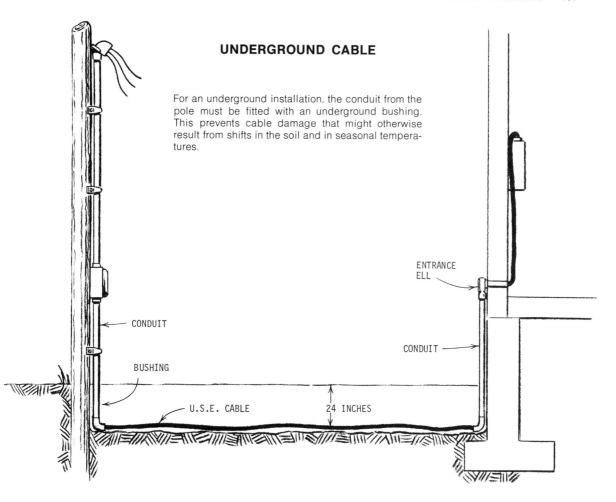

UNDERGROUND CABLE

For an underground installation, the conduit from the pole must be fitted with an underground bushing. This prevents cable damage that might otherwise result from shifts in the soil and in seasonal temperatures.

ENTRANCE ELL

CONDUIT

CONDUIT

BUSHING

U.S.E. CABLE

24 INCHES

U.S.E.

Cable employed underground must bear the "U.S.E." label, which stands for Underground Service Entrance.

Any time you're working on a hot service panel observe all safety precautions, plus:

1. Always pull the main fuse or flip the main circuit breaker.

2. Do not touch the service panel with bare hands. Wear dry gloves and rubber-soled shoes. Stand on a dry board and touch the panel with one hand only. Don't allow your free hand to come in contact with anything such as walls, doors, or appliances.

WIRING A FUSED ENTRANCE PANEL

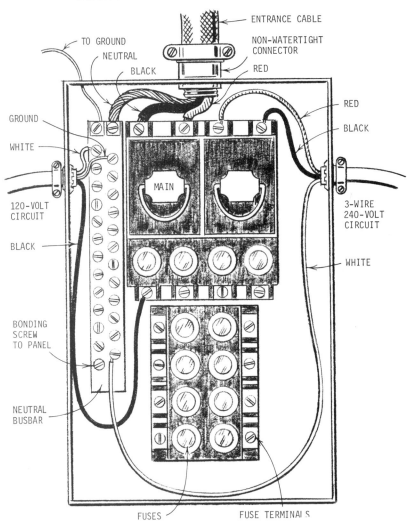

This drawing of a fused panel shows typical entrance wiring at the top, wiring for a 120-volt circuit at left, and wiring for a 240-volt circuit at right.

After you've connected wires into the panel, call in the utility company to install their meter, as shown, and to connect their service wires to yours at the entrance head.

100-AMPERE FUSELESS SERVICE ENTRANCE PANEL

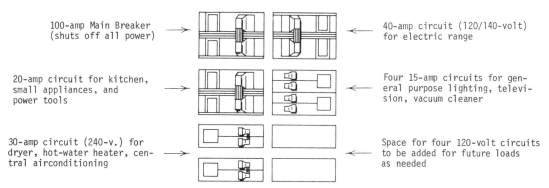

100-amp Main Breaker (shuts off all power) →

20-amp circuit for kitchen, small appliances, and power tools →

30-amp circuit (240-v.) for dryer, hot-water heater, central airconditioning →

← 40-amp circuit (120/140-volt) for electric range

← Four 15-amp circuits for general purpose lighting, television, vacuum cleaner

← Space for four 120-volt circuits to be added for future loads as needed

WIRING A CIRCUIT-BREAKER ENTRANCE PANEL

SERVICE ENTRANCE CABLE

NEUTRAL   RED   GROUND

BLACK

MAIN CIRCUIT BREAKER

120 GROUND

WHITE WIRE

100 ON   100 ON

NEUTRAL BUSBAR

20-AMP SNAP-IN CIRCUIT BREAKER

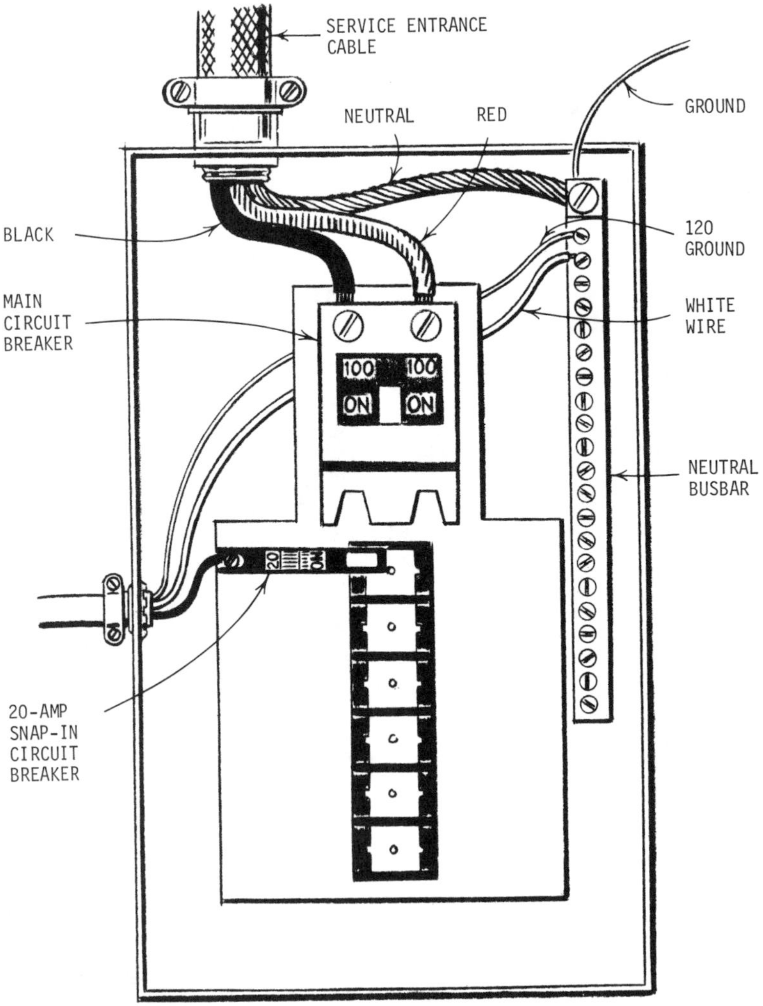

This is a fuseless, circuit-breaker entrance panel. Shown are service entrance connections and those for a single 20-amp circuit.

WIRING A CIRCUIT-BREAKER PANEL

After the service panel has been completely installed, *turn off the main breaker,* punch out the appropriate knockout hole, and bring in a circuit wire or wires. Install a cable connector in the hole. Make sure the wires are long enough to go around the inside of the box to reach the proper terminals. Connect the white wire to the neutral busbar, and the bare ground to the same neutral bar. Connect the black wire to a circuit breaker of the proper amps and snap it in place. Turn back on the main breaker switch.

With the entrance wires in place, *turn off the main* and prepare to bring in cable by knocking out the required number of knockout holes.

Pull enough cable through for the connections and then install the cable connector.

Connect the grounding electrode conductor to the neutral busbar. Insert bonding screw (between neutral bar and panel).

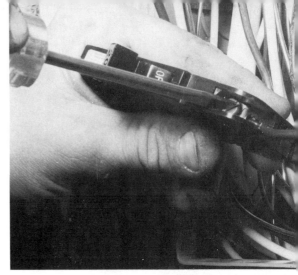

Connect the black to the circuit breaker.

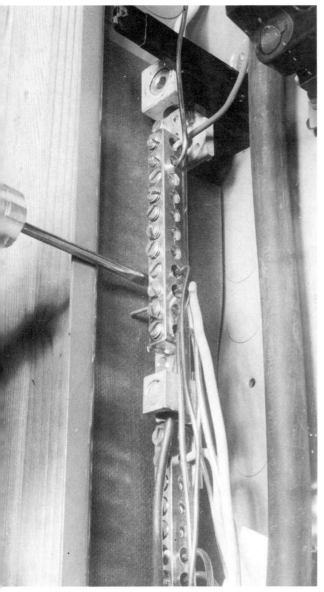

Connect the white to the neutral busbar also.

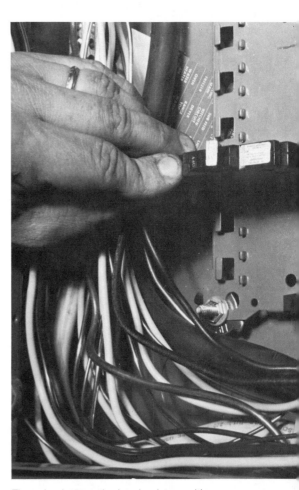

Then simply snap the breaker into position.

CONNECTING CIRCUIT WIRING INTO A FUSE BOX

Again, *remove the main fuse* and connect the white and ground wires to the neutral busbar, which is normally a silver-colored bar with many terminals. Connect the black wire to a terminal connected to a fuse holder. Install the fuse; replace the cover and the main fuse.

Make sure you take a bit of time to properly install the wires in the service panel, running them neatly around the sides of the box rather than here and there randomly in a sloppy manner. Neatness can also promote safety.

GROUND

One of the most important factors in service entrance installation is the ground. Without a *proper* ground, you're worse off than if you have no ground, because you're depending on the ground to save you when actually it's not there.

There are basically two types of grounds: system grounds and equipment grounds.

• In a *system ground,* the neutral incoming wire is grounded, and all neutral wires of a circuit are grounded.

• In *equipment grounding,* all metal parts of the system are grounded. In any metal type of wire raceway such as conduit or greenfield, the metal equipment is used as a ground. However, in the case of nonmetallic plastic-sheathed cable, only the system is grounded, unless you ground each box, fixture and receptacle.

For ranges and clothes dryers, use a 3-wire receptacle with a ground. When you "ground" the appliance frame and boxes, you have a *system and equipment ground* in one ground wire.

One special note of caution: The neutral busbar is not normally grounded in a new service box, and a screw is provided to firmly bond the bar to the box, making an effective ground.

GROUNDING METHODS

The most common method of grounding, and considered the safest, is to run the ground wire to a pipe of an underground cold water system. If at all possible, make the connection on the street side of the water meter so in case the meter is removed for repair, the ground remains connected. If the ground cannot be made on the street side of the water meter, it's wise to install a jumper as shown on page 109; then if the meter connections to the pipe are poor, the ground will still be good. The Code also requires that a metal underground water pipe (used as a ground) be supplemented by an additional electrode, such as a driven ground rod.

The ground should be run the shortest distance to the nearest water pipe. Wire used for a ground must be no smaller than No. 8 copper for a 100-amp service and No. 4 copper for a 200-amp service, and it should be securely fastened in place. In areas where the wire might be damaged, it should be protected by a strip of wood or other relatively nonconductive material.

The ground wire is clamped to the pipe. Make sure your clamp and the pipe are made of the same metal. For instance, on an iron or steel pipe, use an iron or steel clamp. The Code prohibits solder joints on any ground wire connection. In all installations using conduit as an equipment ground, make sure you use proper fittings at all connections.

One of the most common grounds for rural use is the "made ground," or as the Code calls it a "made electrode." The most common made electrode is a copper rod at least ½ inch in diameter and at least 8 feet long. Check with local codes on the grounding rules. The rod should be driven at least

GROUNDING

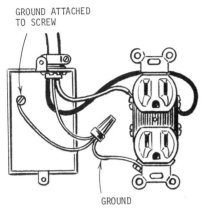

GROUND ATTACHED TO SCREW

GROUND

You can achieve an equipment ground by including a ground wire with all wire runs, connecting all ground wires for a continuous ground, and grounding all boxes and the service entrance.

FOR BARE ARMORED GROUND WIRE

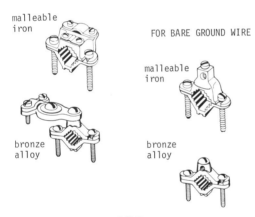

malleable iron

FOR BARE GROUND WIRE

malleable iron

bronze alloy

bronze alloy

FOR RIGID CONDUIT OR E.M.T.

bronze alloy

FOR RIGID CONDUIT

bronze alloy

Here are typical grounding fittings.

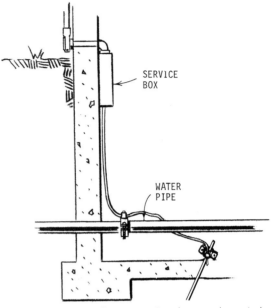

SERVICE BOX

WATER PIPE

This shows grounding to a water pipe supplemented by a "made electrode" such as a ground rod.

All exposed ground wires outside must be protected. A wood strip does an adequate job.

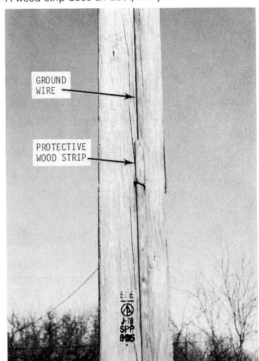

GROUND WIRE

PROTECTIVE WOOD STRIP

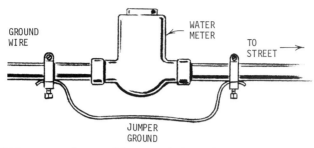

If for some reason you can't connect the ground wire on the street-side of the water meter, use a jumper ground. This precludes the possibility of a poor ground connection through the meter itself.

In rural areas, the Code allows grounding to a driven or drilled well casing, provided the casing is at least 10 feet in length.

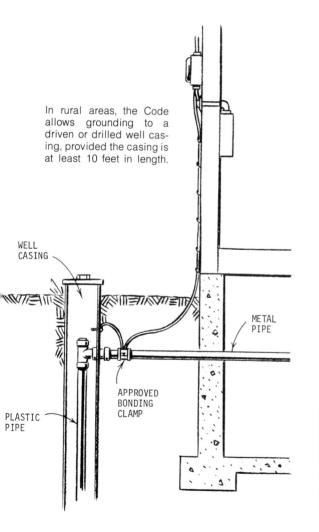

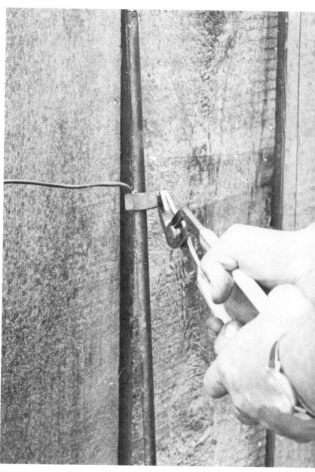

The copper ground wire should be fastened to the ground rod (made electrode) by means of a copper grounding clamp.

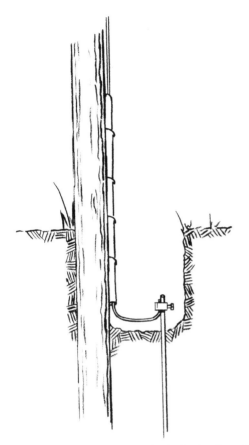

The "made electrode" may be buried below the soil surface, and need not be accessible.

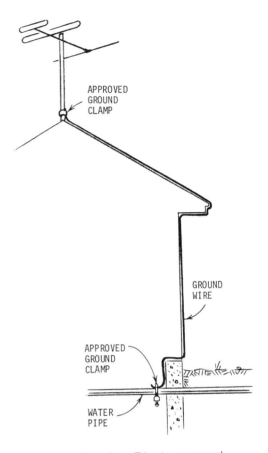

This is a typical run for a TV antenna ground.

8 feet into the ground, and a copper ground wire no larger than No. 6 should be clamped firmly to the rod by means of a copper clamp. Some localities require that the top of the rod be buried at least 2 feet underground as well. Some require the connection to be accessible (or visible).

Once again, installation of a proper ground is of the utmost importance. If at all possible, particularly in rural areas, it's a good idea to have an electrical inspector or someone from the utility company check out the ground installation. The ground's re-sistance may not exceed 25 ohms.

Grounding Television Antennas

It's also a good idea to ground the tele-vision antenna. This is especially important in rural areas where farms are located on high or open locations. In fact the Code re-quires a ground on the antenna. Once again ground as before, using the best possible means.

9

Wiring a New House

In electrical language *new house wiring* means wiring in a house or building that is in the stages of construction, enabling the electrician to make installations before the walls and ceilings are covered. New house wiring may also refer to wiring done in rooms added to an existing house, with the new wires joined to the existing ones.

There are four basic kinds of wiring materials used in new houses today: 1) non-metallic plastic-sheathed cable, 2) thin-wall conduit, 3) flexible metal cable, more commonly called "greenfield," and 4) armored cable. Your preferences, and—more important—local codes determine which type of material and installation you choose. Non-metallic, plastic-sheathed cable is by far the easiest material to install. However, some local codes may require the use of thin-wall conduit or "greenfield," and these two materials are a bit more solid than the others. In most localities thin-wall conduit is required for multiple dwelling units such as duplexes and condominiums.

Assuming the house shell is already up—including outside wall coverings, roof, win-

dows and doors—it's time to install permanent electrical wiring. In most cases power from the utility company is run to a temporary installation that enables the workmen

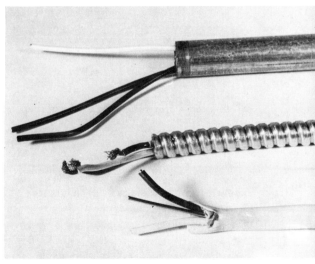

Basic kinds of wiring runs in new houses shown here include conduit (top), armored cable (middle), plastic-sheathed cable (bottom).

to construct the house. The first step is to install the service entrance and circuit breaker or fuse box, as described in Chapter 8.

Make a working drawing that shows where you want each switch receptacle and fixture to be installed. Then make up a plan of the various circuits, so you can effectively run the wires without any waste of materials or time. With this paperwork done, you're ready to begin.

NONMETALLIC PLASTIC-SHEATHED CABLE INSTALLATION

Plastic-sheathed cable is by far the most popular material and the easiest to work with, if your local codes allow it. Chapter 5, on materials, gives you the specific Code requirements for the two different types of nonmetallic sheathed cable. Use type NM for indoor use, and type NMC for indoor or outdoor use.

The cable in new houses must be supported every 4½ feet of run, with additional support within 12 inches of any metal junction, receptacle, switch or light box. Although

Some people use insulated staples to support plastic-sheathed cable. Others use cable straps, like this one, to prevent damage to the outer covering during installation.

some people use staples for anchoring the wire in place, special cable straps are better for this purpose. The straps don't pinch or bend the cable as staples will if they're driven in too far or at a bad angle. Any bends made in the cable must be gradual.

If the cable is pulled down sharply or kinked, you may break the insulation and injure the cable. The Code states that if a bend in plastic-sheathed wire were to be made, that no bend shall have a radius less than 5 times the diameter of the cable.

Step-by-Step Installation

After installing the service entrance, circuit breaker or fuse box, ground, and per-

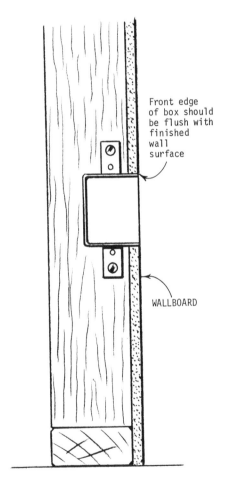

Front edge of box should be flush with finished wall surface

WALLBOARD

All receptacle and switch boxes must be installed flush with the finished wall surface.

haps after hooking up a temporary circuit to run power tools, the next step is to install all boxes. You should know in advance what type of wall covering will be used so that you can place the front edge of the box properly. The Code requires that boxes or fittings installed in walls and ceilings of wood or other combustible material be positioned so that their front edges are flush with the finished surface. On noncombustible walls such as tile and concrete, the boxes and fittings may be installed with the front edges set back not more than ¼ inch from the finished surface. Measure up from the floor to position each receptacle box the same distance from the floor—if over a countertop, the same distance from the countertop. Position all switches at the same height also.

There are several different methods of installing plastic or metal boxes. Some metal boxes on new work are simply nailed into the stud by driving No. 10 nails through the two side holes in the box into the stud. Junction boxes, or light fixture boxes are installed by driving 1- to 1½-inch roofing nails or screws through the holes in the "bottoms" of the boxes and into the ceiling or floor joists. The plastic boxes and some metal boxes are installed by driving ½-inch roofing nails through the metal hanger strips attached to the boxes. Again, it's very important to position the boxes properly, making sure they are spaced correctly and hanging straight. Some boxes have adjustable "hangers" so that you can install the boxes by nailing the hangers to a stud on either side of the box and then slide the box to the right location, without your having to nail the box itself to the stud.

The usual procedure in new house work is to install the boxes and run the wiring from the house power panel to the boxes. *But do not connect to anything in the boxes*

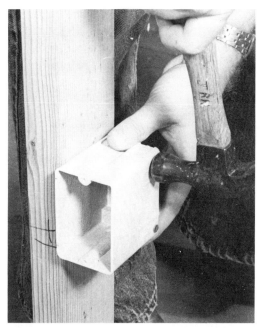

There are many types of boxes and means of fastening. Note that this box is nailed so that it will be flush with the finished wall.

The easiest way to fasten a metal box is to use 1½-inch roofing nails.

BOX INSTALLATION

Here a right-angle bracket is nailed into a joist to support a plastic junction box.

In this installation, the outlet box is positioned against a stud, and then secured with 16 penny nails.

Boxes can be supported on hanger bars nailed to studs or joists. Then you can adjust box position.

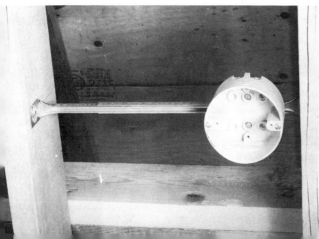

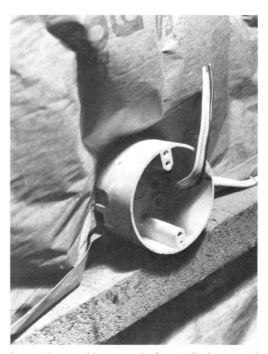

In new house wiring, run all wires to the boxes and then wait with connections until after you've finished the walls.

The Code requires that plaster openings around boxes be sealed. Waiting to mount fixtures until after plastering can save you some cleanup.

In new work, a spade bit on a power drill bores quickly through supports.

or to the house panel. Wrap a piece of masking tape around the end of each wire at the entrance panel and mark which circuit it is, such as "bedroom lights." This makes hooking the wires into the power panel much easier and more orderly.

Since the Code requires that all openings around boxes be sealed, you should hold off on further electrical work until the wall covering has been installed and the openings have been plastered. Plastering is a messy job. So you'll be glad that you have only the boxes to clean of plaster before proceeding. (Sometimes when installing wall paneling in tight places, it's best to leave the box unsupported, then cut the paneling so the box can fit in it with the ears of the box fitting over the paneling. Fasten the box securely with screws and/or box clamps.

Running wiring in new work is actually fun when compared to the chore of rewiring old work. It's merely a matter of bringing the wiring to each junction box, then to each receptacle and switch. Normally you'll have very little cutting and notching to do in the floors and ceilings, but you will have to bore holes in some places and through all the walls. One method of boring is to use a paddle or spade bit in an electric drill. This type of bit cuts fast in the soft woods used for house construction. Naturally, if you don't have electrical power to the building, you'll have to bore with a brace and bit. In this case, use a ratchet style brace so you can bore close to walls and corners.

When you have to run through holes bored in floor or ceiling joists, or through wall studs, always drill the holes in the center of the joists. This is a safeguard against

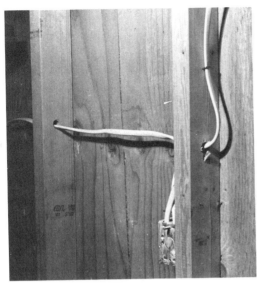

This is a typical installation of receptacles along a new-house, plastic-sheathed "cable run."

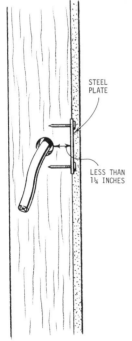

If the cable will pass within 1 1/4" inches of the nearest edge of the stud, nail on a protective steel plate.

STEEL PLATE

LESS THAN 1 1/4 INCHES

a nail's being driven into the wiring later. If the holes are located less than 1¼ inches from the nearest edge of the stud, the front of the wooden surface should be covered with a metal plate or bushing of at least 1/16-inch material to protect against nails being driven into the wire years later.

If the cable runs parallel to a ceiling or floor joist, it may be stapled to one side—one staple at least every 4½ feet. If the cable runs through the joists, again you'll have to bore holes. Cable should never be left unsupported for long spans. When cable must be run down to a second floor, basement or up in an attic, on new work it's merely a matter of drilling up or down through the floor or ceiling plate *inside* the unfinished wall and running the wires through.

In using this type of material for running wires, no extra protection such as porcelain tubes, etc., is required when you run the wires through the holes. An alternate method of running wires across the joists in an attic or basement is to nail support boards across the joists, and fasten the wiring to these boards. If an attic is accessible by a drop ladder or stairs, the wires running at angles or across the joists must be supported on either side by guard strips at least as high as the wiring. If the attic has a crawl opening, only an area 6 feet around the opening must be protected in this manner. In all instances the cable should follow as closely as possible the contour of the building construction. It should never droop across areas unsupported.

As you pull a new wire into a box, pull it through at least 8 inches and bend over the wires so they can't be pulled back out of the boxes. Once the walls have been covered, the wallboard installed and plastered, connect each fixture into the wiring, making sure all fixtures and boxes are grounded.

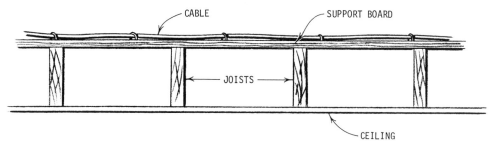

When a wire spans the spaces between joists, it's wise to mount it on a support board.

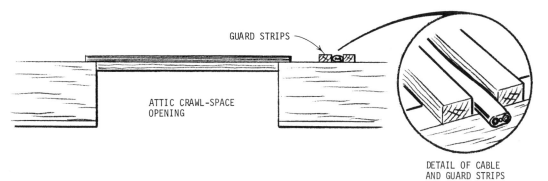

If an attic is accessible, all exposed wiring on the attic deck must be protected by guard strips that are at least as thick as the wiring.

Then connect up all junction boxes to complete the circuit, install the junction box (if used) covers, and finally connect each coded wire from each circuit into the circuit breaker or fuse box. *Make sure the mains are pulled so no current is in the terminals, and follow all safety rules on service entrances, outlined in Chapter 8.*

ARMORED CABLE

Armored cable is required by some local codes, and it is installed in much the same manner as NM-type (nonmetallic) sheathed cable. Like nonmetallic sheathed cable, ar-

mored cable can only be used in permanently dry locations. Running armored wires through studs and joists is handled in exactly the same manner as described for plastic-sheathed wires except that you use cable staples. (Armored sheathing is not as apt to be crimped and damaged by the staples as plastic sheathing is.)

Greenfield is a flexible metal conduit, similar in shape to armored cable. However, greenfield cable does not come with the wires already installed. With greenfield, you run the empty cable into position and fasten it to each box. Then before the walls are com-

CONNECTING ARMORED CABLE

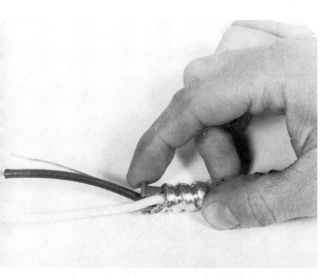

Armored cable comes prewired. It is installed with methods much like those used for plastic-sheathed cable. To fasten it to a box, cut off the outer sheathing, leaving wire exposed. Then slip on a plastic bushing, as shown.

Run the cable into the box and turn down the cable-clamp screw to hold the cable in place. Cable should extend a minimum of six inches into the box.

Prepare the box by turning down the inside ring on the box's locknut.

pleted, you pull the wires through using fish tape. Either armored or greenfield flexible metal cable is used in areas where thin-wall conduit won't make the needed bends.

CONDUIT

Thin-wall conduit is being installed in a great number of today's homes. It is commonly termed "EMT" (electrical metal tubing). Though thin-wall conduit takes a bit more work and some special tools for installation, some local codes deem it the safest and require it in all new work. Consult your local power company or electrical inspector.

Conduit is run in much the same manner as the various cables, but because of its

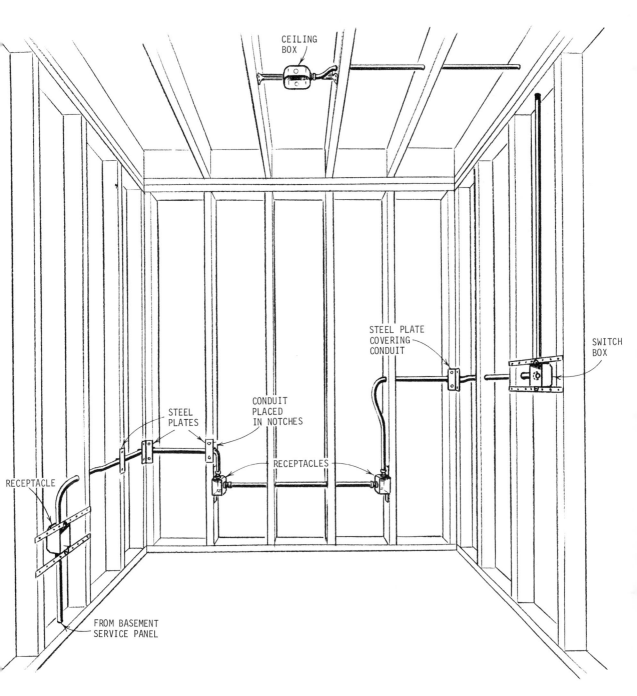

CEILING BOX

STEEL PLATE COVERING CONDUIT

SWITCH BOX

STEEL PLATES

CONDUIT PLACED IN NOTCHES

RECEPTACLES

RECEPTACLE

FROM BASEMENT SERVICE PANEL

Local code may require that you use thin-wall conduit, commonly termed E.M.T. This illustration shows various box mounts and the protective steel plates.

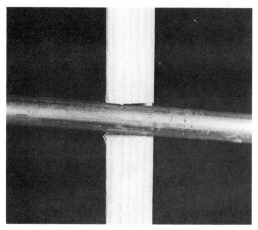

Conduit can be run through holes bored in supports or into notches, as shown here. Conduit in a notch should be protected by a steel plate.

semi-rigid qualities, it cannot be as simply installed. Conduit must be bent with a special conduit bender. It is then cut, fitted and connected to boxes and to other pieces of conduit with special couplings. Conduit is normally placed in notches cut in the wall studs. Then a steel strap is nailed over the stud to prevent later accidental nailing through the conduit.

For installation, you must first mount all boxes. Then you run the conduit to the boxes and make hookups using special connectors. Next you pull the wires through the conduit with fish tape and make splices for future connecting to the fixtures, receptacles, and switches. Remember to leave at least 8

These are common support straps for conduit. The upper left-hand drawing shows a protective steel plate.

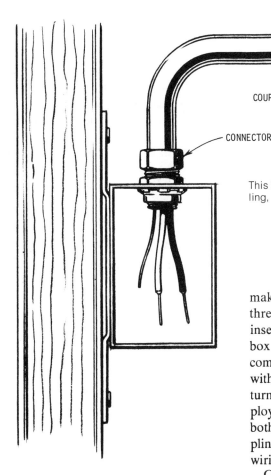

This is a typical conduit connection, showing a coupling, a connector and a box.

inches of wire protruding from each box to connect to the fixtures. Last, you connect wires in the breaker or fuse box.

Working with conduit takes some special techniques, a great deal of patience, and at least two people—one to push the wires through, and one to pull and work them into position. Only steel boxes are used. Conduit comes in 10-foot sections which you join with couplings. And it should be supported about every 10 feet and within 3 feet of each outlet box.

Thin-wall tubing is never threaded. To

make connections, you insert it into the threadless end of a special fitting. Then you insert the threaded end of the fitting into a box knockout and tighten the locknut. The compression type of fitting consists of a body with a split ring which is forced in place by turning the nut. Another type of fitting employs a special indenter tool to deeply indent both the fitting and conduit, but this coupling is more commonly used on commercial wiring. There is also a set-screw type.

Conduit may be cut with a hacksaw (using a blade with 32 teeth to the inch) or with a common pipe cutter. In any case, the end must be reamed to remove any burrs which might damage insulation on the wiring as it is pulled into place.

CONNECTORS

COUPLINGS

Here are conduit couplings and connectors.

1. CAP OFF, WIRE STRAIGHT THROUGH

2. BEND MADE

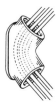

3. WIRES PULLED, CAP REPLACED

After running the conduit to boxes, you can pull the individual wires into place. A pulling elbow, like the one shown, helps get the wires through bends.

With conduit, wires must be continuous from outlet to outlet. Do not splice wires together, because the wires could then pull apart or short out. Make sure any wires you pull through the conduit conform to regular wiring standards. That is, in a 2-wire circuit, you need one black and one white. But in a 3-wire circuit, you need one black, one white, and one red. The metal conduit serves as the equipment ground.

One of the Code rules says that you must avoid making more than four full quarter bends in a run of conduit, from one outlet to another. And you must avoid abrupt bends which might make it difficult to run wires through the conduit.

The basic wiring of a new house is quite easy and well within the capabilities of the average handyman. What's more, if you've done your own wiring and trouble arises later, you know your whole wiring system and can quickly determine the problem.

10

Rewiring an Old House

In comparison to all other types of wiring jobs, rewiring an older home is the biggest challenge. Because the work is hidden and hard-to-get-at, this type of wiring is the hardest and can be frustrating. It often takes a great deal of ingenuity to run wires through walls, but a few tricks of the trade and some special tools can make the job a lot easier.

In rewiring an older home there are actually two distinct types of work: 1) repairs or additions, and 2) complete rewiring. The first step in wiring an older home is to make a thorough check to determine if the wiring is adequate and—more importantly—safe! In some cases the wiring, even if old, may be safe enough to use but inadequate for today's increased demands for electricity. Adding a new circuit breaker or fuse box of the proper amps plus installing the additional circuits needed may make the wiring adequate.

Wiring is one of the most common home fire hazards, and this is particularly so in areas with lax code rules. Old wiring is usually to blame in electrical fires. So in an extremely old home, unless the wiring has recently been redone, you can bet it's a distinct fire hazard.

You can normally date the wiring by the type of materials used. The oldest form of wiring is the old "knob-and-tube" wiring in which porcelain knobs were used as anchors to fasten individual wires to the framework of the building. Porcelain tubes were used as "insulators" for passageways bored in the holes in the wood, and the wiring is run through these. Although some people may consider this type of wiring adequate, many areas do not allow additional outlets to this type of wiring, so it should be replaced. A second and somewhat later style of wiring utilizes a flexible cable similar to the style used now, except the insulation on the wiring is different from the plastic sheathing used today. One sure sign of bad wiring is broken and cracked insulation. If you spot any bad places in the old wiring, you can bet there are more, probably hidden from view.

Another common hazard with older wiring is wire splices which are not enclosed in a metal or plastic box. Again this may be a fire hazard.

Old wiring is a major cause of house fires. If the insulation is cracked and frayed, replace it.

Another danger in older homes is ungrounded wiring. Some people recommend that ungrounded receptacles in older homes be grounded by running an individual ground wire from a receptacle to the nearest water pipe or other approved ground. However, the cost and time involved in this operation are as much or more than running a completely new and properly grounded system throughout the house. The main rule in rewiring is to run grounded wire to everything and make sure all grounding rules are followed. This is covered in chapters 6 and 8.

It's also a good idea to contact the local electrical inspector and ask what his ideas are on the rewiring project. You may be required to submit a plan of work, which he will either okay or revise. Then he must inspect and okay the work after you're finished. Many people feel this is a nuisance and a personal insult. But a good electrical inspector may locate hazards that could

otherwise cost you your home and the lives of your family. Get to know him well, and "pick his brain," because he's gained a lot of experience in the field. Most inspectors are quite willing to offer advice.

Treat the rewiring of an older home just like the wiring of a new home. Determine your wiring needs and make a good plan for the placement of receptacles, switches and lights. Then draw a diagram or plan of the best routes for running the wires. This plan is sometimes difficult to make for older houses, especially two-story houses and houses with odd-angle roof lines which make it impossible to run wires from the attic down through portions of the outside walls. You'll use a lot more material in older work because you normally can't employ the straight-line approach of new work. Instead, you'll have to go around obstacles and through walls. The idea is to install the wiring without damaging any more of the building than necessary.

Running wiring in any type of dwelling already finished is considered "old" work and is covered under Code rules as such. Naturally, it's not possible to use rigid-conduit or tubing without tearing out walls. So the most commonly used wiring is plastic-sheathed cable or armored cable. However, some local codes require flexible metal tubing called "greenfield."

When completely rewiring an older home, you will almost always need to replace the old fuse box with a circuit breaker box or a new larger-amp fuse box. Install this first, as well as a new service entrance. And have the electricity hooked up to it by the utility company. In cases of complete rewiring, none of the old wiring is connected to this new box.

It's a good idea to completely wire a circuit. Then, as the last step, connect it into the circuit breaker box. You can connect a partially wired circuit to the box, as long as all exposed wires are thoroughly wrapped with electrician's tape and there are no areas that might cause shorts. For a complete rewiring job, you might as well count on doing without electricity for a few days, at least until you can get one circuit installed. Some power companies will allow you to install a temporary circuit at the box, so that you can use power tools. Before installing additional circuits, make absolutely sure the entire house power is shut off by pulling the *main* fuses or *main* circuit breaker.

There is just no way each individual problem in old wiring can be explained, because each is very different. But there are certain "tricks" and methods that make it easier to get a wire from one point to another.

One method that saves a lot of time and trouble is to utilize the old wiring. Discon-nect the old wires from any switches, lights and receptacles, and then remove the old hardware. With the old wire in place, splice it to the end of the new wire. Then, as you pull out the old wire, you'll pull the new wire into place. Make sure the splice is tight and well wrapped with plastic electrician's tape to insure that it doesn't come undone or catch on obstacles.

Pulling and fishing wire in older work is a two-man operation. It is most easily done with one person feeding the wire into the hole, and another gently pulling it out another hole. Otherwise, the wire has a tendency to kink up. In feeding wires down a wall you can often push in twice the amount of wire you actually need unless someone is below to pull it out.

Another method of pulling new wires through walls and ceilings employs electrician's fish tape. This is a flexible metal strip

Here you see frayed, cracked, and crumbled insulation on old wires—a sure sign of danger.

It's often possible to tape new wire to old and thus pull new wire into place while removing old.

one end is extremely effective for snagging a wire that has been run down a wall to a point near a receptacle opening.

In older homes where redecorating will be done anyway, the easiest method of running wire from one point to another is to cut a hole in the wall wherever studs and joists present obstructions. Then you chisel chan-

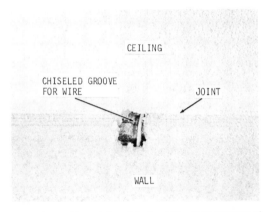

Sometimes it's not feasible to drill passageways at places such as ceiling plates. In this case, open the wall or ceiling, chisel a channel, and replaster.

Rewiring an older house takes ingenuity and a second person to help push and pull wires. Try to keep all twists out of Type NM cable if possible.

that can easily be pushed through passageways. First you push and "work" the tape through the route. Then connect the tape to the wire and "fish" the tape back along the route, thereby pulling the wire into place. A homemade "fishing" device consisting of a piece of coathanger wire with a loop on

nels for the wire into the obstructions. This chiseling helps especially at the joint between the ceiling and wall. A wire can be easily fed down from an attic into a wall by bringing it around the ceiling plate, instead of by boring through the plate. This method is particularly useful on outside walls where it is almost impossible to get to the plates to bore holes. This works also under baseboard or ceiling moldings. Just remove the moldings; chisel a channel in the plaster; run the wire and reinstall the molding. Make sure the channel is chiseled in a straight run and that you nail the molding back in place on either side of the wire so there is no danger of driving a nail into the wire.

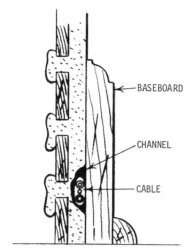

Channels can also be chiseled behind baseboards, as shown here, or behind ceiling moldings.

A long extension bit or extension for a regular bit and brace is almost a necessity for drilling through wall plates, and at the odd angles required to make runs through ceilings and walls.

There are normally two areas to which most of the wiring is run in an older home. You can install the circuit junction boxes in the attic, or in the basement. If the home is a single story with an unfinished attic, the most practical approach is to run the wiring through the attic. In a two-story home you may need to run the wiring in the basement, and in some cases, the attic as well.

When installing attic wiring, it is a simple matter to lift the attic floor boards and run the wires under them. If the attic is completely unfinished, you can simply fasten the wires to the ceiling joists.

When running wires through an unfinished basement, bring the wires down through a wall, then run across the basement and back up through the other walls. Make sure to support the cable with cable straps on all joists.

Two-story homes provide the most challenge because the wiring must be fished from either the attic or the basement and then into the ceiling for fixtures.

You may need an extension bit for boring through ceiling plates and through wall cross braces.

The most important secret in fishing wires is to place the openings in line so it's simpler to bring the wires through the holes. In almost all cases the second floor wall will lie directly over the lower floor wall, and this necessitates drilling a hole. (See drawings on upcoming pages.) The first step is to remove the baseboard from the second floor and drill down through the floor plate and the ceiling plate. Bring the wire down from the attic using a fishing wire and out a channel chiseled at floor level and behind the baseboard. The wire can then be brought

COMING DOWN FROM THE SECOND STORY

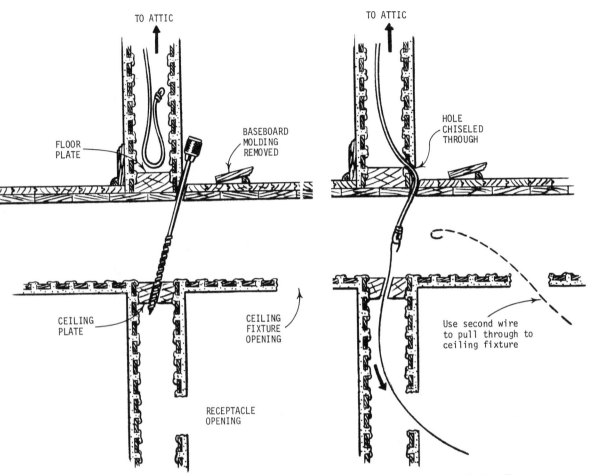

The left-hand drawing shows how to bore a passageway from the second story. Then using a flexible metal wire, pull the cable through the passage and out the wall opening. If you also wish to bring the cable to a ceiling box, catch the cable with a second fish wire; as shown. Note: Baseboard was removed.

down to a switch or receptacle in the wall below. You can also bring a wire down from the second floor and pull it across to a ceiling fixture by using a second fishing wire. Again this type of operation is best done with two people.

You can use a similar technique for bring-

ing a wire up from a basement and into an outside wall for a receptacle or switch. The same principle can be used when bringing the wire down through the attic into the basement; the two drawings on this page show the baseboard removed and the way in which a channel is cut for the wire.

Again, a main goal when running wiring in older homes is to do a quality electrical job while damaging the building as little as possible. I guarantee you'll run into some mighty hard problems, but with a little time and thought almost all problems can be solved.

If you plan to completely redecorate with perhaps new wallboard or paneling, installation becomes much more simple. For instance, it allows you to notch around studs and run the wires from one side of a room to the other. You can then cover the exposed wiring with the new wallboard or paneling. It's a good idea to cover the exposed wiring with a piece of thin sheet metal to insure that no nails are driven into the wiring during installation of paneling, or at some time later.

One way to save on costs of wires is to place receptacles and wall lights back to back on the sides of a wall. You can then run wire between the two units.

A good orderly procedure is to first run the main lines to the junction boxes either in the attic or basement. Then drop the wires

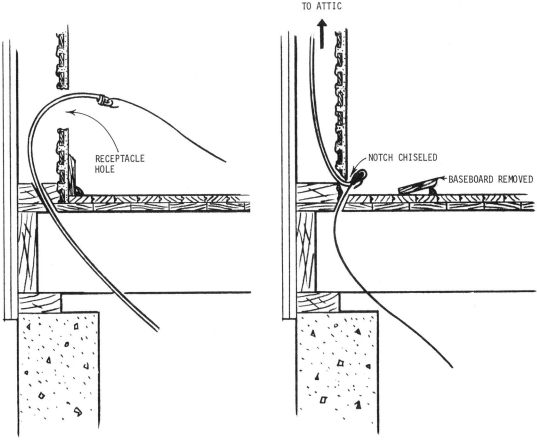

Here are common methods of bringing wire up from the basement and down to the basement from the attic.

down through the walls, or bring them up from the basement to the switches, receptacles or lights.

After running the wiring to the places where boxes will go, you install the boxes and connect the wires to the fixtures. After all this is completed, connect the circuit into the circuit breaker box or fuse box and test the circuit to make sure there are no problems. Chapter 16 describes how to test circuits and locate specific problems.

INSTALLING BOXES

In some instances you may wish to use existing holes for installation of switches and receptacles, but you may also wish to change them. And you will definitely need to install new receptacles because very few older homes have the required number of receptacles per room.

In the case of using an existing opening, you've probably already removed the fixture, outlet or switch to pull new wires through. Then it's merely a matter of connecting a new box in place and wiring the device.

Adding new receptacles presents a bit more of a problem. For good workmanship all receptacles should be placed the same distance from the floor, and all switches should be evenly placed, if possible. Measure up for the positioning.

Using your knuckles or a hammer, tap on the wall to determine if any studs would interfere with your intended location of the box. If the location seems to be away from a stud, try to cut the hole, causing as little damage as possible. In a lath-and-plaster wall, use a chisel to tap a hole through the wall to determine the location of the laths. Then using the box as a pattern, or a piece of cardboard cut to the right shape, mark the wall for the box opening, and cut the opening using a keyhole saw. Make additional notches at the top and bottom center to allow clearance for the receptacle or switch screws.

With a saber or keyhole saw, cut out the opening, cutting down on each side of the center path. In the case of a plaster wall, the center lath should be cut out first, while the upper and lower laths should be cut only partially to provide a solid surface for fastening the box. After making partial cuts in the upper and lower laths, snap lath out by inserting a screwdriver and carefully twisting. The main problem in cutting holes in lath-and-plaster walls is that the plaster crumbles from around the hole while you saw the laths. One trick is to use a hacksaw blade with one end taped, and cut only on the pull stroke, holding the plaster with the opposite hand. With the opening cut, try fitting the box inside. You may have to do additional cutting for a snug fit.

When you're satisfied that the box fits properly, remove it. Then, using a coathanger wire, fish out the new wire that you have already dropped down behind the wall. Bring out the wire and insert it in the box, leaving about 6 inches protruding. If you're using boxes with built-in clamps, screw the clamp down tight on the wire. If you're using boxes with knockout holes, punch out the hole and place a wire clamp in position. Fasten the box in place.

In most instances, if you have correctly fitted the box in place it can be secured to the laths. Some boxes are equipped with sliding, reversible ears which allow you to fasten the box in place on either wallboard, paneling or lath and plaster. The main consideration is that the front of the box rests less than ¼ inch from the edge of the opening. Preferably, it should be flush with the wall surface. If the laths have vibrated loose and the plaster is crumbled and

INSTALLING A RECEPTACLE IN LATH-AND-PLASTER

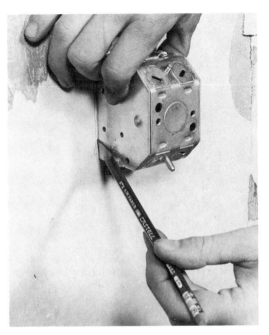

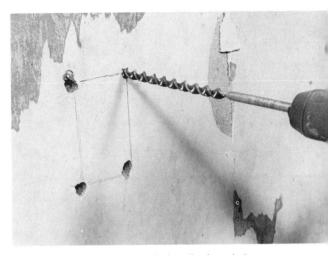

2. On dry wall or wood-paneled walls, bore holes at all four corners of the box outline, as shown. This permits entry and turns of a keyhole saw.

1. Before marking the intended opening by tracing the box outline, as shown, use a stud finder or knock on the wall to be sure the cut won't hit a stud.

3. On lath-and-plaster walls, first cut a small hole (approximately where you would like the center of the box) to determine exact lath positions. Then use the box as a template to trace an outline that will result in lath positioning as shown. Cut plaster away with a chisel. Bore holes in lath at box corners to permit use of a keyhole saw.

5. Check the box for fit. To get a snug fit you may have to trim the laths a bit.

6. After running the cable into position inside the wall, fish it out with a bent coathanger.

7. Pull the wire into the box, screw-fasten the receptacle to the laths, and plaster around the opening.

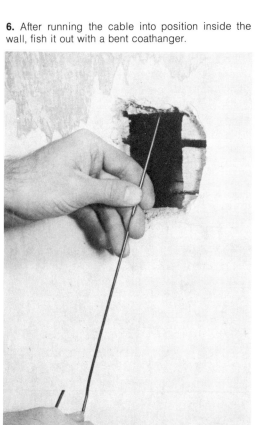

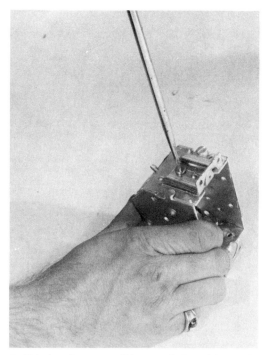

8. This box has reversible ears that enable you to fasten the box to either wallboard or lath.

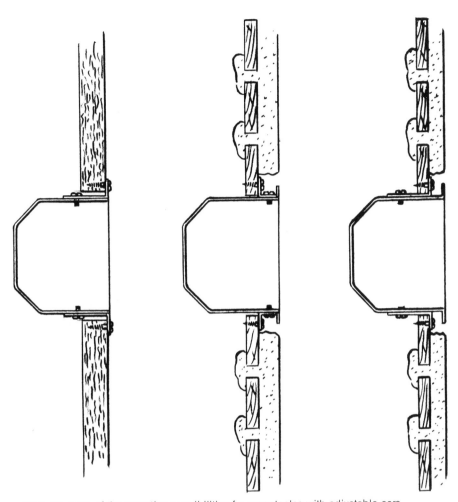

Here are some of the mounting possibilities for receptacles with adjustable ears.

broken, you may have to resort to using the special box shown on page 134. This box is placed in position and the screws on the side flanges are tightened to secure the box.

Another method of securing a loose box is to use box hanger strips. These light metal strips are slipped in behind the wall on each side of the box, and the metal ears are bent inside the box, securing it in place. Use a thin pair of pliers and squeeze the ears completely back against the sides of the box to insure there will be no contact with these ears and the terminals of the receptacle or switch.

The last step after securing the box is to replaster around the opening. The Code requires that all plaster be repaired so there are no gaps or open spaces around the boxes.

After all wiring has been run and the boxes installed, the receptacles, switches and light fixtures can be installed.

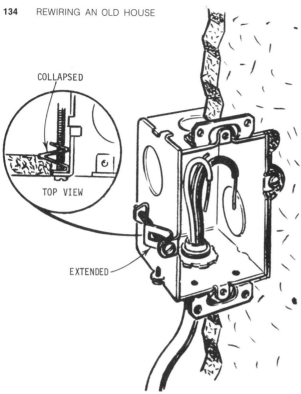

COLLAPSED

TOP VIEW

EXTENDED

On this special metal box, the side flange anchors the box against the inside of the wall when you turn down the flange screw.

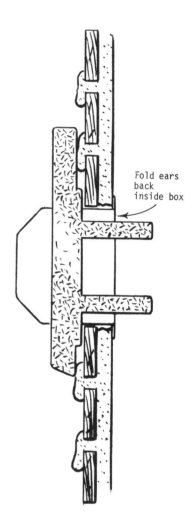

Fold ears back inside box

If it's not feasible to use a box with screws or with a side flange, you can use a special box hanger strip, shown at left and above.

To finish the job, simply wire-in the receptacle, mount it on the box, and test-install the cover as shown to ensure it fits flush with the wall. If plaster is cracked as shown, either repair it or install an oversize cover.

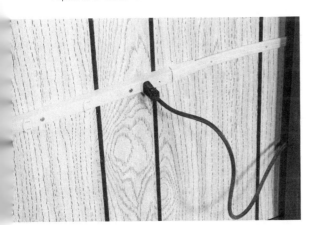

This surface-mounted receptacle strip offers an alternative to punching holes through walls in order to gain more outlets.

When rewiring older houses it is sometimes necessary to use some special materials. For instance if you wish to install a ceiling light in a lath-and-plaster ceiling you have two alternatives. If the light fixture falls between joists, and is in a one-story house, the simplest method is to cut the opening in the ceiling and use a fixture outlet box with an attached hanger strip. These hanger strips will extend to fit between the studs and are nailed into the studs at each end of the hanger and provide support for the box. Again the opening must be replastered to insure a smooth finish around the box.

If a deep box would damage the building

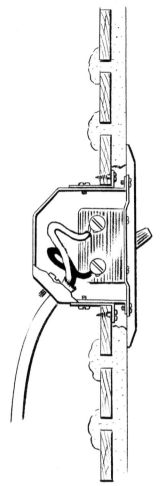

This is a cutaway view of a completely installed switch in lath-and-plaster.

construction, you can use a ½-inch-deep shallow box.

Both shallow boxes and the normal size boxes can also be attached to a lath-and-plaster ceiling in a two-story house by means of a "hanger bar." This is a metal bar with a threaded stud in its middle. The hole for the box is cut in the ceiling, the bar pushed up in the hole and placed across the lath-and-plaster. The box knockout hole in the center is punched out and fitted over the threaded stud. Then a nut is turned down to pull the bar tight against the lath-and-plaster and to draw the box up inside the opening.

TROUBLESHOOTING

If the wiring in an older house continually shorts out, causing blown fuses or tripped circuit breakers, turn off the current at the specific circuit breaker or fuse causing the problem. Unplug all appliances and turn off all lights in the circuit. Then turn the circut breaker on again or replace the fuse. Plug in the appliances and turn on the lights, one by one, until you find the troublemaker, but do not use an oversize fuse. If a receptacle is the problem, the box or the receptacle itself may be loose due to someone yanking an extension cord out. The terminal shorting against the sides of the box. In this case it's merely a matter of anchoring the box securely and repairing or replacing the receptacle. If the circuit does not trip the circuit breaker or blow a fuse immediately, but does so after some time, the circuit is overloaded. Some of the appliances need to be put on a new circuit, or perhaps a new separate circuit is needed.

In very few cases will it be possible to run a branch line to extend a circuit in an older home because most older circuits are already overloaded.

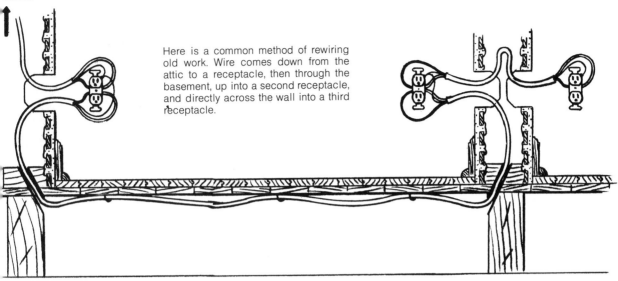

Here is a common method of rewiring old work. Wire comes down from the attic to a receptacle, then through the basement, up into a second receptacle, and directly across the wall into a third receptacle.

BASEMENT

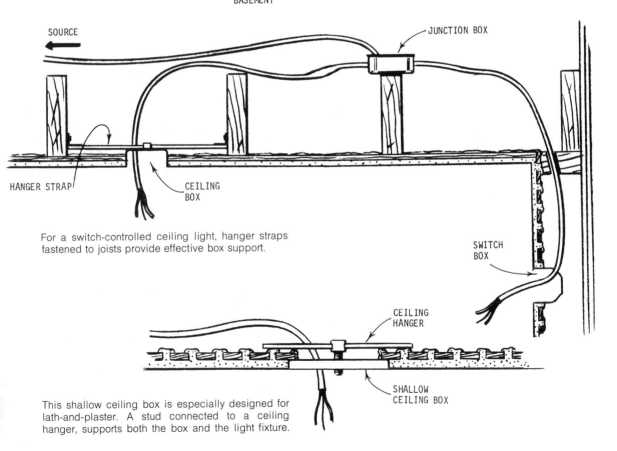

SOURCE

JUNCTION BOX

HANGER STRAP

CEILING BOX

SWITCH BOX

For a switch-controlled ceiling light, hanger straps fastened to joists provide effective box support.

CEILING HANGER

SHALLOW CEILING BOX

This shallow ceiling box is especially designed for lath-and-plaster. A stud connected to a ceiling hanger, supports both the box and the light fixture.

ADDING NEW CIRCUITS

In a home with fairly modern and grounded wiring, but with an inadequate number of circuits, additional circuits may easily be installed if the breaker or fuse panel will handle them.

The first step is to run the wires for the new circuit, starting at the breaker or fuse box, *but do not connect them to the box.* Again run the wires as described at the start of the chapter, cut the holes for the boxes and install the receptacles, switches and

When bringing new circuits into a new fuse box or circuit breaker box, first place a cable connector in the appropriate knockout hole, as above.

lights. When the circuit has been completed, remove the panel from the front of the circuit or breaker box. Pull the *main* circuit breaker or fuse to shut off the entire house power. *Use all the safety precautions mentioned in Chapter 8, on service entrance wiring, when wiring new circuits into breaker boxes.*

Carefully remove the knockout hole in the side of the box. Strip the cable and insert it through a cable clamp installed in the box.

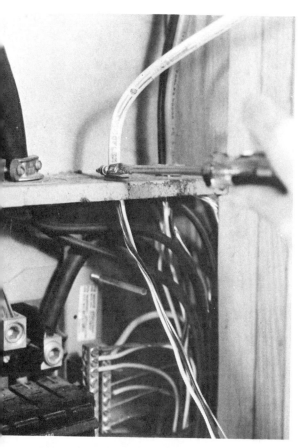

Strip new wire ends and bring them through the cable connector hole, as shown at left. Then secure the cable connector shown above. (Note: This series of photos was shot *before* electricity was connected to the service entrance. So the *main* breakers are shown in place. Normally *you should pull main fuses or turn off main circuit breakers before you begin work.*)

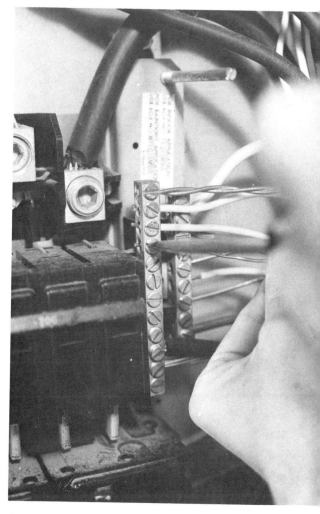

Connect white and ground wires to the neutral busbar.

If the box contains fuses, connect the black wire to the terminal screw on the fuse holder, and connect the white wire to the neutral terminal bus. Connect the grounding wire to the neutral terminal bus. Place a fuse of the proper amperage in the circuit. Then replace the cover and perform a test to insure that it is working properly.

If the box contains circuit breakers, fasten the white wire and the grounding wire to the neutral terminal bar. Then fasten the black wire to a circuit breaker of the proper amperage and snap it in place. Flip the *main* circuit breaker, and again test the circuit.

Connect the black wire to a brass terminal on the circuit breaker unit and snap the unit into place. Were this a fuse, you'd connect the black wire to the appropriate terminal and then insert the fuse.

Then test the circuit.

11

Wiring for Heavy-Duty Appliances

The Code classifies appliances in three major categories:

1. Portables include light appliances such as toasters and coffee percolators. Normally these require just 120 volts.

2. Stationary appliances include window airconditioners, self-contained ranges, and clothes dryers. Normally these require 240 volts and are often equipped with a plug and cord—and may be moved about fairly easily.

3. Fixed appliances have to be installed permanently and in compliance with the Code. Among these are water heaters, countertop cooking units, and built-in ranges, ovens and garbage disposals. Although some of these fixed appliances may be connected by means of a simple plug and cord, they are still considered fixed appliances, according to the Code.

Local codes are extremely restrictive about how a 240-volt circuit may be installed. So make sure you check with your utility company and local wiring inspector as to the proper materials and methods. Naturally your service must be modern, consisting of a 3-wire service. And it also must be adequate to carry the load required by the appliances.

In general, it's a good idea to wire a separate circuit to each of these appliances individually:

1. Water heater
2. Range
3. Separate countertop and oven
4. Garbage disposal
5. Clothes dryer
6. Dishwasher
7. Any motor or other fixed appliance operating at over 12 amps
8. Any fixed appliance operating at 240 volts
9. Any motor that starts automatically, such as a furnace and an electric heating baseboard
10. Any permanent, and some of the larger, window-type airconditioners.

The wires used for these appliances should also be heavier than the common No. 14 or 12 used in ordinary house wiring. No. 10 wire is considered a minimum by code in

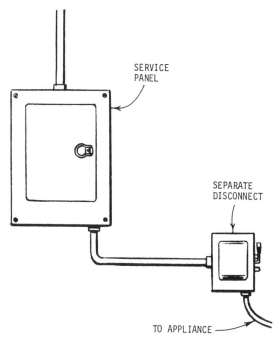

SERVICE
PANEL

SEPARATE
DISCONNECT

TO APPLIANCE

"Fixed" appliances, those that cannot be moved easily, should have a special disconnect between the appliance hookup and the service panel. A circuit breaker may serve as a disconnecting means.

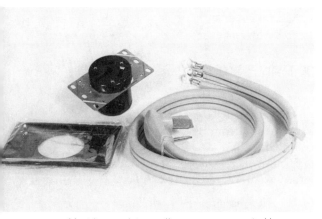

Most heavy-duty appliances are connected by means of a special receptacle and a pigtail plug and cord.

most instances; No. 6 and No. 8 are commonly used.

There are basically two ways of wiring these circuits. With one, the wiring is run directly from the appliance to the service entrance panel. But if the appliance is "fixed" and cannot be easily moved, the Code requires a means of disconnecting it, sometimes achieved with a separate disconnect box, shown in the accompanying drawing. However the most common disconnect device is a heavy-duty receptacle, connected to the appliance by an appliance cord, shown in the photo. For ranges and countertop units, the cord has a "pigtail" type connector.

The receptacles used for heavy-duty, 240-volt wiring and pigtail installation differ in configuration from the plug slots of ordinary household receptacles. This is to prevent the possibility of someone accidentally plugging a 120-volt appliance into a 240-volt circuit. There are literally hundreds of different configurations. Some of the receptacles have nonlocking plugs. Others are of the locking type, which means that the plug is first inserted into the receptacle and then turned to make it operational. Before it can be removed, it must be turned back to its "insert" position.

The receptacles are identified as "2-pole 2-wire" or "2-pole 3-wire," or even "3-pole 4-wire." The number of poles listed is the number of conductors normally carrying the current. In the event there are more wires indicated than poles, the receptacle has an extra connection for the use of a separate grounding wire. This terminal or connector should be used only for a grounding wire, and never for a wire that would be carrying current under normal conditions.

The tiny figures marked on the illustrations indicate where the neutral (**W**) and ground (**G**) are located.

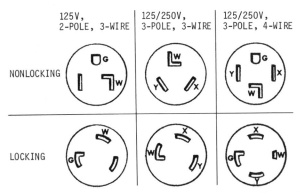

There are many configurations for mating plugs and receptacles. These shown are for 30-amp hookups. Note the general difference between locking and nonlocking designs.

The receptacles may be flush-mounted with the wall surface, or they may be heavy-duty wall receptacles that protrude from the wall.

Each appliance circuit must also be equipped with some means of completely disconnecting it from the circuit for repairs or cleaning. Naturally, on portable appliances, disconnection is done by merely unplugging the appliance. This also goes for some stationary and fixed appliances. However, if the appliance is rated at 300 watts or less, no special equipment is needed. Also, if the rating is higher but the circuit is protected by a circuit breaker, no further disconnect means or pullout fuse block is needed. If the circuit is served by a plug-type fuse, a completely separate disconnect must be installed, as was shown on page 142. If the appliance is "fixed" and equipped with a cord and plug, or an easily accessible, receptacle-pigtail device which may be unplugged, you may use a branch circuit switch as a disconnect.

The Code is very strict as to the type of overcurrent protection which must be provided for appliance circuits. Any circuit which serves a single appliance and does not serve anything else may use an overcurrent protection device of up to 150 percent of the rated amps of the appliance.

Remember, in electrical wiring the white wire in a cable must be used only as a neutral. Using 2-wire cable for wiring a 240-volt appliance, you must paint the white wire black at both ends of the connection to indicate that the wire is actually a "hot" wire. Although the Code does not place a restriction on the type of wiring to be used for wir-

The receptacle at left is commonly used for air-conditioners. The other one is a surface-mounted type. Be sure that receptacle capacities match the requirements of the appliance.

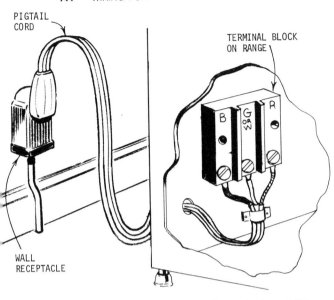

PIGTAIL CORD

TERMINAL BLOCK ON RANGE

WALL RECEPTACLE

Here are a typical wall-mounted receptacle and the connections for an electric range.

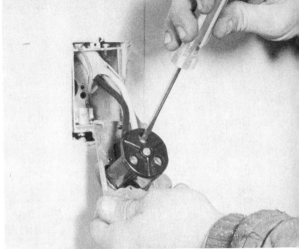

Use the same techniques for wiring heavy-duty receptacles that you would for regular ones, but follow the connection instructions on the backside. Then mount the receptacle and the cover.

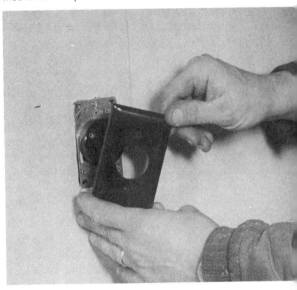

ing appliances, many local codes do. So check thoroughly. In some setups where *the cable runs directly from the overcurrent protection device in the service panel,* 3-wire service entrance cable is used to wire ranges, cooking tops, ovens, or dryers.

WIRING SELF-CONTAINED RANGES

The range operates at 240 volts and utilizes No. 6 cable in a 50-amp circuit. To install the cable, connect the circuit to a 50-amp, 240-volt circuit breaker or fuse in the service panel. Note the connection of the pigtail on the range. The proper wire connection will be marked on the back of the range-connection terminal block. The black wire of the pigtail is connected to the "B"; the red wire is connected to the "R"; and the white or green wire is connected to the "W" or "G." Note: The metal frame of the range must be grounded to the neutral terminal. This does not normally require a sep-

arate wire because most of today's modern ranges are constructed so the frame is automatically grounded by the neutral wire when a pigtail cord and range receptacle are used. In all cases the neutral wire must be No. 10 or larger.

Many of today's modern kitchens have a separately installed countertop cooking surface and an oven or ovens instead of the single self-contained range. In some cases these can be installed on a single circuit. However, it is normally better to put them on separate circuits.

If you install separate circuits, the oven should be installed the same way as shown for the self-contained range on page 144. However, the wire size will normally be somewhat smaller, probably No. 10. Again, nothing smaller than No. 10 can be used because of the need for grounding. Normally a 30-amp receptacle is suitable for an "oven-only" installation. For a fixed or installed oven, which cannot easily be removed, the pigtail-to-receptacle arrangement cannot serve as a disconnect means. So unless the circuit is protected by a pullout fuse block, or a circuit breaker, you must have a sepa-rate disconnect.

To install the countertop cooking oven, proceed in exactly the same manner, again preferably using a No. 10 wire and a recep-tacle and pigtail installation. When installing both the countertop unit and the oven on the same circuit, make sure the circuit is large enough to carry the load. To de-termine this, add together the ratings of the two separate units to determine the wire size and amperage of the circuit. Then proceed as if the two units were one.

One important note: The neutral to any oven, self-contained range, or cooking top must be insulated unless you use service entrance cable, and originate at the main service panel.

There are two methods of acquiring the 240 volts needed for the appliances. The

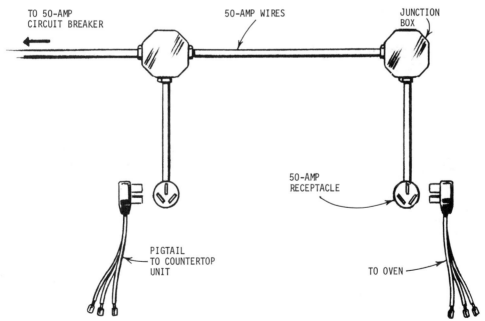

This 50-amp, 240-volt circuit powers a countertop cooking unit and a separate wall-mounted oven.

most common method used in a service panel which has a "range" circuit breaker or pullout fuse block is to merely connect to terminals of that specific device. However, this type of installation will serve only as one circuit. To acquire the additional 240 circuits needed, wire the black and red wires to a 50-amp, 240 circuit breaker. The white wire goes to the neutral busbar again. To wire heavy-duty 240 circuits to a fuse box, connect the white wire to the busbar and the black and red wires to the proper fuse terminals, according to the wiring diagram on the inside of the box cover.

ELECTRIC CLOTHES DRYERS

Electric clothes dryers are normally installed by the same methods used for the range. Smaller size wire can be used, in most cases a No. 10. Install a 30-amp receptacle and pigtail installation. Some local codes may require the use of a pullout fuse or circuit breaker in the main service panel *plus* an additional safety disconnect switch between the service panel and the dryer.

Some of the newer dryers are called high-speed dryers. They normally operate at 8500 watts. These require a heavier circuit,

Here, pulled for display, is a plug-type fuse with connection for a 240-volt range circuit.

If your service panel doesn't have a main range fuse or circuit breaker, you can install one, as shown.

normally a 50-ampere circuit and a heavy-duty receptacle to match.

CLOTHES WASHERS

Most codes aren't specific on the installation of washers because some are connected to the plumbing by merely screwing on rubber hoses, and others by securing to metal pipes. This means they can be classified as either stationary or fixed. In most cases the washer is cord-and-plug equipped so that connection is merely a matter of plugging

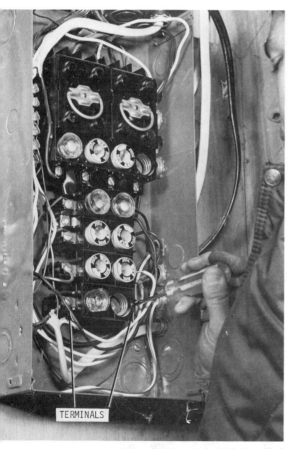

Here is how an additional 240-volt circuit is installed on the entrance panel.

into a receptacle. The frame of the washer will normally be grounded, *if the receptacle is properly grounded.* An added safety precaution would be to install a GFCI on the circuit.

WATER HEATERS

There are two types of water heaters: single element and double element. The double-element type of heater is the more commonly used today because it gives a constant supply of hot water. It has two thermostats, whereas the single-element heater has only one. The size of the wire and amperage of the circuit vary with local codes. So check with your local code or inspector as to what installation may be needed. Some require a No. 10 installation, others a No. 12.

Be very careful to provide the proper grounding. If your service equipment is a circuit breaker panel, connect the heater using a 2-pole, 20-amp breaker, and No. 12 wire (if local code allows). If you use No. 10 wire, provide a 2-pole, 30-amp breaker. The installation in fused equipment is somewhat different. If the service panel does not have an unfused tap especially constructed for the water heater, install a separate safety disconnect as shown on page 148.

In some areas the power companies will offer a lower rate for the current used for water heaters if they are equipped with a special "current peak load device." This actually switches the water heater off during periods when the utility company has high power demands. In this case the time switch is installed, and the wires are installed the same as before, except that they must start at the utility company time switch, instead of at the normal service panel.

AIRCONDITIONERS

Some smaller airconditioners are merely plugged into a 120-volt circuit. However,

WIRING A WATER HEATER

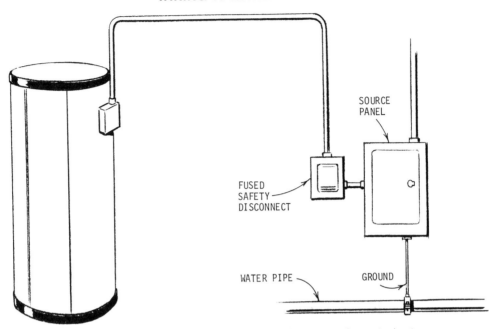

Most local codes require a separate means of disconnecting water heaters.

After bringing plastic-sheathed cable into the water heater junction box, wrap the water heater leads around the feeder wires and then secure them in place with wire nuts.

Finally, attach the grounding wire to the grounding screw on the terminal block, or if need be, to the metal cabinet.

some of the larger ones, even window style units, must be supplied by 240. Normally these are merely plugged into a heavy-duty receptacle utilizing at least a No. 10 wire, 30-amp circuit. If the unit has a 3-phase motor and operates at over 250 volts, it must be installed on a circuit of its own and use a special disconnecting means as discussed on previous heavy-duty appliances.

FURNACES

Most furnaces will operate on 120-volt electricity, and are merely connected in as a normal circuit. However, the furnace should be on its own circuit and utilize at least No. 14 wire.

GARBAGE DISPOSALS

Most garbage disposals are merely plugged into a normal receptacle which is installed in the wall under the sink. They could be on a 15-amp circuit with at least No. 14 wire. A wire is run from the receptacle to a

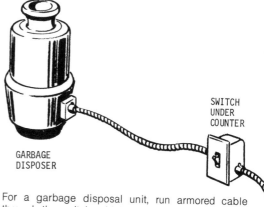

GARBAGE DISPOSER

SWITCH UNDER COUNTER

For a garbage disposal unit, run armored cable through the switch, which may be located above or below the counter.

switch above or below the counter which allows the disposal to be turned on and off. The main concern in installing these appliances is that they be properly grounded.

ELECTRIC BASEBOARD HEATING

Today more and more people are using electric baseboard heating, especially in rooms that have been added on to an existing home. Some baseboard heaters are installed using 120 volts, others using 240 volts. Check with your local codes as to what is allowed, then purchase the proper unit. Check the manufacturer's installation instructions and follow them carefully.

ELECTRIC BASEBOARD HEAT

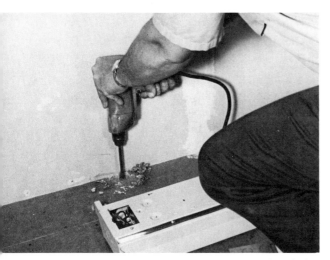

1. Electric baseboard heating is a popular choice for add-on rooms. Drill the hole for cable.

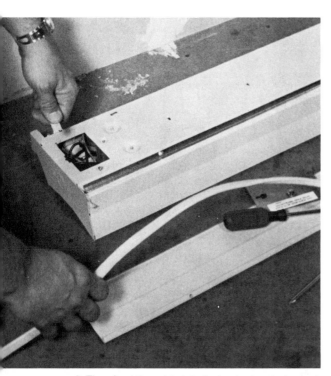

2. Then feed the cable into the heater terminal panel.

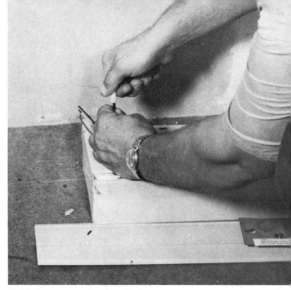

3. Make connections to the heater wires.

4. Fasten the heater to the wall.

5. And mount the cover strap in place.

GROUNDING

Of primary importance in all heavy-duty wiring is that the system be thoroughly grounded. Using the 3-wire and pigtail, as described earlier, the equipment is probably grounded properly, *if the appliance is properly constructed and the circuit is grounded.* If the appliance is permanently installed and wired using armored cable or conduit which is anchored solidly to the terminal block on the appliance, it's probably grounded properly. If you are using nonmetallic plastic-sheathed cable, a ground wire must be attached to the frame of the appliance, and also to the nearest box. When you are using nonmetallic cable, the bare conductor (unlike SE cable) can be used for "grounding" purposes only. The neutral conductor must be insulated.

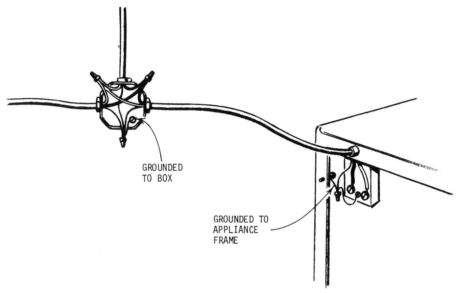

GROUNDED
TO BOX

GROUNDED TO
APPLIANCE
FRAME

In heavy-duty wiring, be sure you properly ground the system. In this drawing, note the grounds to the terminal block and to the appliance's metal cabinet and also the ground in the nearest metal junction box.

12

Shop Wiring

Most do-it-yourselfers have a shop. It may only be a small workbench located in one corner of the garage, or it may be a complete shop occupying most of a basement or even a separate building. Receptacles in garages must be protected by ground-fault circuit interrupters (GFCIs). More and more these days, shops are being designed and outfitted for the entire family. They may be set up for wood and metal work, for auto repairs, and for all kinds of family craftwork.

There are three important factors to consider in wiring a home shop: *proper lighting, adequate power* for the electrical tools, and—most important—*proper grounding* of all tools.

The floor plan on the next page shows where lights and special electrical items would be installed in a complete shop. You will note that there are two rooms separate from the shop. One is a finishing room, which is a must if you plan to do much furniture finishing. One of the biggest dangers in home shop work is spraying finishing materials such as lacquer in a basement near a gas furnace. The spray room doesn't have

to be very large, but it should be equipped so it can be sealed off from the rest of the shop. And it should be equipped with an explosion-proof exhaust fan and an explosion-proof light—that is, fixtures that won't ignite dust or vapors.

The second, extra room shown is a craft room. It can be used for all sorts of handicrafts. A separate craft room is essential for keeping workshop dust off the crafts.

LIGHTING

In a shop you need plenty of lighting for safety. There are two methods of lighting a shop: 1) flooding the entire room with light by installing banks of fluorescent fixtures across the room and 2) installing one or two main lights, as well as special lights at each work area or tool. The choice is a matter of preference.

In any event, the shop should have at least one or two main lights which can be turned on when you enter the room. If the shop is in a basement or a separate building you will probably wish to install 3-way switches on the main lights so you can turn

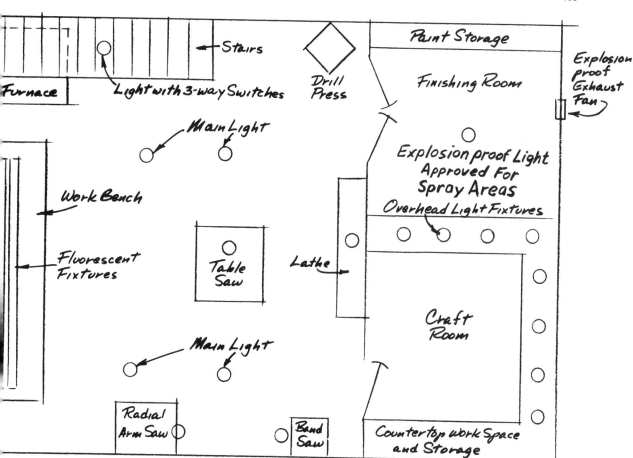

Stairs

Light with 3-way Switches

Furnace

Paint Storage

Drill Press

Finishing Room

Explosion proof Exhaust Fan

Main Light

Explosion proof Light Approved For Spray Areas

Overhead Light Fixtures

Work Bench

Fluorescent Fixtures

Table Saw

Lathe

Craft Room

Main Light

Radial Arm Saw

Band Saw

Countertop work Space and Storage

A good workshop wiring plan provides for plenty of receptacles and shadow-free illumination of work surfaces.

In this large shop, banks of ceiling lights provide a uniform brightness throughout.

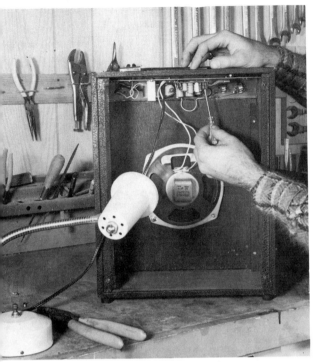

A portable high-intensity lamp for the workbench helps afford optimum illumination.

the lights on before entering the shop, and turn them off when you leave. If you are using the individual light system, each working area should have its own light, and each light should be properly situated for the specific tool or work area. The light should be located so it does not glare in your eyes. Yet it should provide the right amount of light at the most important part of the work surface.

A workbench should be equipped with a set of fluorescent fixtures or incandescent lights situated no more than two feet apart, for the entire length of the bench. This floods the workbench area with light. When working on small items such as electronic devices, you'll need additional light from a specific angle. A small, portable, high-intensity light is just the ticket for that.

Again, in a workshop, each tool must be lighted properly. For a radial arm saw, the light should not be positioned exactly over the top of the saw because this would cause the saw arm to shade the cutting area. Locate the light slightly to the left side. On the other hand, a light positioned exactly over a table saw blade will enable you to see the work clearly.

Lighting a wood lathe, I like to situate the light fairly low, pointing it straight down on the lathe. I also use a reflector on the light that throws a sort of "spot" on the wood and gives a good clear view of the profile of the turning wood. Don't situate the light so low that a piece of flying wood could strike it. Also, make sure the light is guarded by some sort of shield. In fact, it's a good idea to use a shield on all workshop lights.

Some tools such as drill presses are better off with a small "spot" type of light mounted directly on the tool. This spot type with a goose neck is also a good means of lighting a sharpening tool such as a belt grinder.

If the light is situated optimally, it will pick up the reflection of the "new edge" as it is ground on the tool. Any major work areas, such as a forge or an electronics bench, should also be lighted for the specific job, and lighted well. Don't skimp on the size of bulbs. Use nothing smaller than 100-watt bulbs on the close-up tool lights and 300-watt bulbs on the overhead lights, as well as for the main lights.

ADEQUATE WIRING

Nothing is more frustrating than to have to unplug one tool in order to plug in another. Make sure you have plenty of receptacles situated around the room. Because the workbench is normally the main work area, you'll find a strip plug worth installing there. You may even wish to install a strip plug

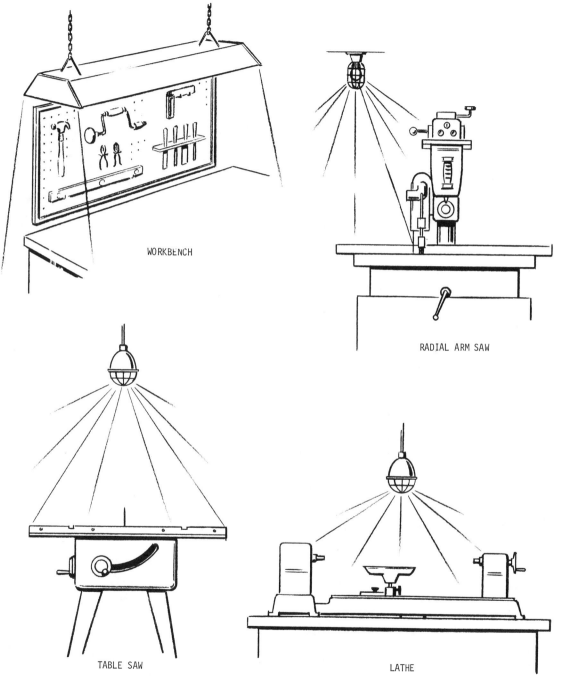

WORKBENCH

RADIAL ARM SAW

TABLE SAW

LATHE

These light positions eliminate shadows on the work. Metal guards over fixtures help prevent bulb breakage.

This shield is designed for a bare bulb in a socket. Note the socket clamp and the top hook.

A gooseneck lamp with base clamp can be adjusted to suit the structure of almost any tool.

On big tools, lights are often standard equipment.

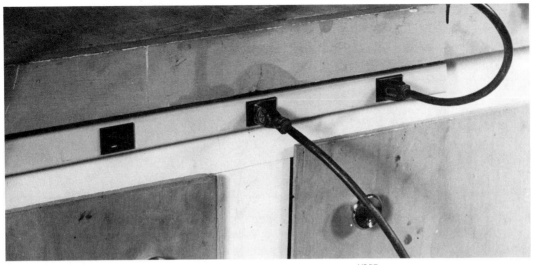

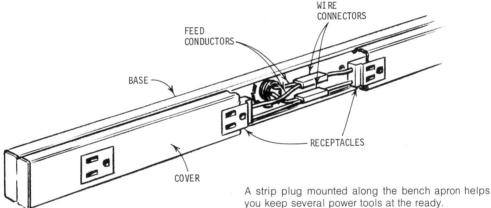

FEED CONDUCTORS

WIRE CONNECTORS

BASE

RECEPTACLES

COVER

A strip plug mounted along the bench apron helps you keep several power tools at the ready.

around the entire room.

There are several different methods of installing receptacles in a workshop. If the shop is fancy and finished in paneling, you will want to install the receptacles in the normal manner, placing them no more than 4 feet apart. If your shop is strictly utilitarian, you will probably use open wiring and surface-mounted receptacles. If you're building a garage shop with no danger of flooding, you should consider installing floor

receptacles. These are installed by using conduit runs in the concrete floor. The boxes are installed at the locations you will be placing tools, and this type of installation makes a much neater and safer shop because you won't have long cords running across the floor.

In addition to the normal receptacles, if you have a heavy-duty motor, such as a 5 horse for an electric planer, you may wish to install a 240 circuit. In this case, you

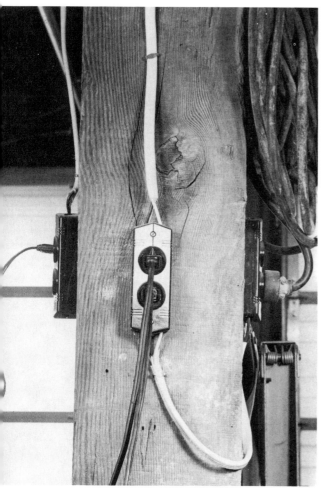

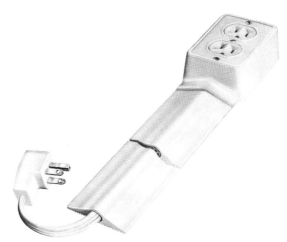

This floor cord with tapered cover provides a sensible alternative to the concrete implacement. And it's much safer than snaking standard cords about.

For heavy motors like the one on this planer, you must install a separate disconnect between the motor and the service panel.

If you don't require a "finished" look, open wiring and surface-mounted fixtures serve the purpose. For permanent wiring, the cable insulation should be 1/4 inch inside of receptacle box.

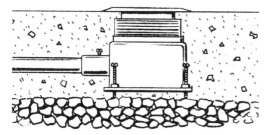

In a shop not subject to flooding, receptacles can be set in a concrete floor.

CIRCUIT WIRE SIZES FOR INDIVIDUAL SINGLE-PHASE MOTORS

Horsepower of Motor	Volts	Approximate Starting Current Amperes	Approximate Full Load Current Amperes		LENGTH OF RUN IN FEET from Main Switch to Motor							
				Feet	25	50	75	100	150	200	300	400
¼	120	20	5	Wire Size	14	14	14	12	10	10	8	6
⅓	120	20	5.5	Wire Size	14	14	14	12	10	8	6	6
½	120	22	7	Wire Size	14	14	12	12	10	8	6	6
¾	120	28	9.5	Wire Size	14	12	12	10	8	6	4	4
¼	240	10	2.5	Wire Size	14	14	14	14	14	14	12	12
⅓	240	10	3	Wire Size	14	14	14	14	14	14	12	10
½	240	11	3.5	Wire Size	14	14	14	14	14	12	12	10
¾	240	14	4.7	Wire Size	14	14	14	14	14	12	10	10
1	240	16	5.5	Wire Size	14	14	14	14	14	12	10	10
1½	240	22	7.6	Wire Size	14	14	14	14	12	10	8	8
2	240	30	10	Wire Size	14	14	14	12	10	10	8	6
3	240	42	14	Wire Size	14	12	12	12	10	8	6	6
5	240	69	23	Wire Size	10	10	10	8	8	6	4	4
7½	240	100	34	Wire Size	8	8	8	8	6	4	2	2
10	240	130	43	Wire Size	6	6	6	6	4	4	2	1

Chart by Sears, Roebuck and Co.

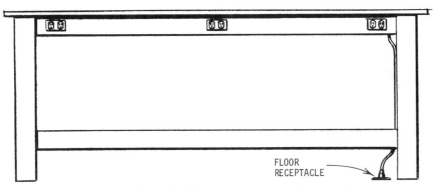

A useful workshop aid, this "hot table" features receptacles all around the apron.

should have a separate disconnect. And be sure the circuit is adequate for individual motors, as shown in the chart at the top of this page.

One of the most important items in my shop is a "hot table." This is a 4 x 8 foot table that stands in the center of the room and is wired with receptacles on all four sides. One end of the table fits against the table saw, and the saw is plugged into one of the table receptacles. The large size of the table enables me to support long boards and full-size pieces of plywood while I rip them over the table saw.

GROUNDING

The most important thing is that all electrical units be grounded. All receptacles must be grounded, and it's a good idea to install a ground-fault circuit interrupter (GFCI) on the receptacle circuits to insure even more safety. Faulty portable electrical tools have accounted for many deaths and injuries, and working on a construction job, I had the misfortune to witness one of these tool-caused fatalities. Today this doesn't have to be a danger, thanks to the use of double insulation on power tools. Be sure that any power tool you purchase in the future is double insulated, to provide maximum protection.

Today, however, the tools themselves are not as much to blame as the operators. One of the most common and dangerous ways to misuse a tool is to break off the grounding plug so it will fit into an extension cord which does not have a grounding slot. This can result in a severe shock. It makes more sense to use an approved extension cord with a grounding conductor.

Wired and lighted properly, a home workshop can be a joy. But set up sloppily and with little forethought, a shop can be a nuisance to work in—and downright dangerous as well.

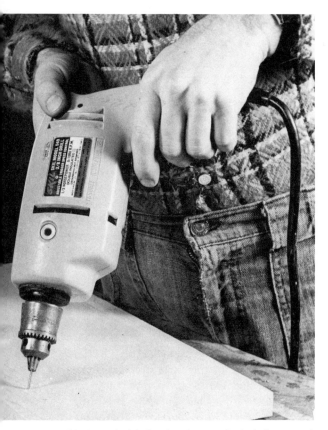

Modern, double-insulated power tools help prevent dangerous shocks.

Caution: Do not break off the third (grounding) plug, as shown, just to gain access to a 2-prong receptacle. Instead, buy an adapter with a ground wire. or better yet, install a grounding-type receptacle.

13

Farm and Ranch Wiring

Farm and ranch wiring is more "involved" than most urban residential wiring because there are more buildings in the system and because nearly all rural spreads have their own water supplies, powered by electric water pumps.

And to keep expenses down, as a farmer or rancher you must become a jack-of-all-trades. This includes being an electrician. When a pump—or even the thermostat or humidity control of a grain silo—goes on the blink, you'll probably have to fix it yourself. But your two major causes of trouble in the wiring system will probably be 1) poor grounds and 2) lightning and storms.

An adequate farm or ranch today requires a 200-amp service as a minimum, which is usually supplied by a single distribution pole. Ideally this pole is centrally located among the buildings, yet within a reasonable distance of the nearest utility company pole. If you don't already have a centrally located "yard pole," the local utility company will normally help you decide where to put one. They may also install it and run power to it and to the meter, al-

though you'll have to install the entrance head or cable. However, if you are located some distance from the nearest utility-company power line, you will probably have to pay for installation of poles up to your yard pole. In most modern installations the pole will also have a main disconnect box as well as the meter connected to the pole. Feeder lines will run from the pole to the house, barns, and outbuildings. At one time it was normal procedure to run the feeder lines to the house, then run additional feeder lines from the house to the buildings. However, the farther the lines extended from the meter, the more the voltage dropped. So the most distant buildings were obtaining insufficient voltage. A better choice is to locate the pole centrally and run separate feeder lines to each building.

Feeder lines may be overhead or underground cable, depending on your preference, ease of installation, and local codes. Overhead cables are more commonly used because of their ease of installation and lower cost. But they present an unsightly appearance; in addition, heavy windstorms and ice can bring these lines down. Overhead lines are especially troublesome when you try to move large equipment under them. So many modern farms now have underground feeder cables.

With 200-amp service located on the pole, the best method of running feeder cables is to run a separate 3-wire feeder to every major building. Run a 3-wire to a 100-amp service entrance (without meter) in the house, a 3-wire to a 60-amp service panel

Often in rural areas the utility company's pole, shown in foreground, feeds the owner's distribution pole, shown in lower left.

Each feeder line into a separate building should have its own separate service box, such as the 30-amp fuse box shown.

in the barn, and a 3-wire to a 30-amp service panel in other buildings such as a brooder house.

An alternate method, which some larger ranches are using, includes separate meters for the house and the most important other building. Although this setup is somewhat expensive initially and is normally frowned on by utility companies because of the

added time involved in reading the meters, it lets you know how much power your "major" business is taking.

If the outbuildings are small and closely grouped, and you plan to use only lights and perhaps one receptacle in each, you can run a 3-wire cable from building to building.

With overhead wires, the main thing is that they be strong enough to withstand

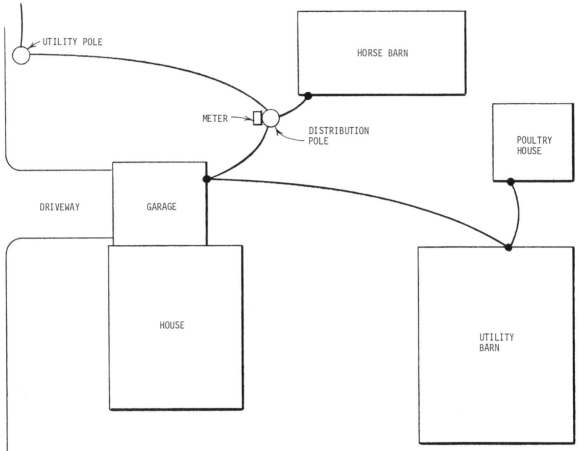

Ideally, the distribution pole is centrally located. If it carries a meter, the pole should allow the utility company easy access for meter readings.

Some local codes allow you to run feeder wires between buildings, provided the buildings are small and closely grouped.

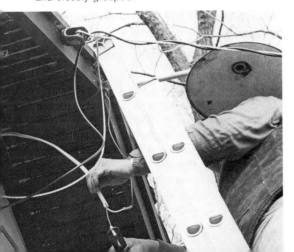

wind and ice storms, as well as carry enough current without a substantial voltage drop. The Code states that on lines carrying 600 volts, nothing smaller than No. 10 wire should be used for spans up to 50 feet; and No. 8 should be used for longer spans. When lines are to carry more than 600 volts, No. 6 should be used for individual conductors; No. 8 is okay in cable.

All feeder wires running to or from buildings must be kept 3 feet from windows, doors, porches, and other openings includ-

Adequate Wire Sizes for Weatherproof Copper Wire

Load in Building Amperes	Distance in Feet from Pole to Building	*Recommended Size of Feeder Wire for job
Up to 25 amperes, 120 volts	Up to 50 feet	No. 10
	50 to 80 feet	No. 8
	80 to 125 feet	No. 6
20 to 30 amperes, 240 volts	Up to 80 feet	No. 10
	80 to 125 feet	No. 8
	125 to 200 feet	No. 6
	200 to 350 feet	No. 4
30 to 50 amperes, 240 volts	Up to 80 feet	No. 8
	80 to 125 feet	No. 6
	125 to 200 feet	No. 4
	200 to 300 feet	No. 2
	300 to 400 feet	No. 1

* These sizes are recommended to reduce "voltage drop" to a minimum. (Chart by Sears, Roebuck and Co.)

ing fire escapes. To determine the proper size feeder wire for each building, you must know the distance from the yard pole to the building and the approximate amperage to be used in the building at one time. (To find amperage, divide the wattage by the voltage.)

DISTRIBUTION POLE (YARD POLE)

As mentioned, the utility company may install the yard pole, or they may not, depending on local codes. But assuming the pole is in position, the next step is to wire it in. Some local codes allow you to install the hardware on the pole before setting the pole in the ground. Then you merely connect the wires to the pole after the pole is set.

When the utility company runs the wires to the pole, the top wire is normally the neutral wire, but you should ask the utility company to make sure. In previous years it was common practice to bring each wire to the pole and mount it to an insulator bracket or to individual insulators. Today, where permitted, the three wires are sometimes braided together, and the neutral wire is bare. But regardless, in a proper installation the neutral wire is normally con-

nected at the top of the pole. In the newer installations the single strand of 3 wires is held in place by a "tension" clamp screwed to the pole. The "combined wire" system greatly lessens the danger of direct shorts, which often occurred when wires of the old 3-wire scheme were blown together during windstorms. In any case, it's a good idea to check local codes before installing the feeder lines. The distribution yard pole contains the power lines from the utility company, the meter, and a switch to turn off the entire unit in case you wish to install more feeder lines.

There are basically two methods of installing the incoming and outgoing lines on the pole. Local codes will determine which may be used. In one setup called a "single-stack" construction, a metal service head connected to a large conduit is strapped to the pole by means of large conduit anchors. (Some localities allow the use of plastic piping.) This conduit is then brought down and installed to a meter base which is normally located about 5 feet off the ground, so that the utility company man can easily read the meter. Another short piece of conduit runs down to the weatherproof circuit breaker box. And the three individual wires, including the neutral and the two hot wires, are run down through the meter head into the service entrance.

The two hot wires are brought back up through the conduit and out the entrance head—with about 36 inches protruding so they can be connected to the various feeders. The neutral wire going down to the meter socket ends there but is spliced into all the feeder lines at the top of the pole.

One thing to remember about installing a service head and conduit is that the top of the service head must be above the insulators or line connectors so there is enough of a drip loop to prevent water from running down the wires into the service head.

SINGLE-STACK YARD POLE

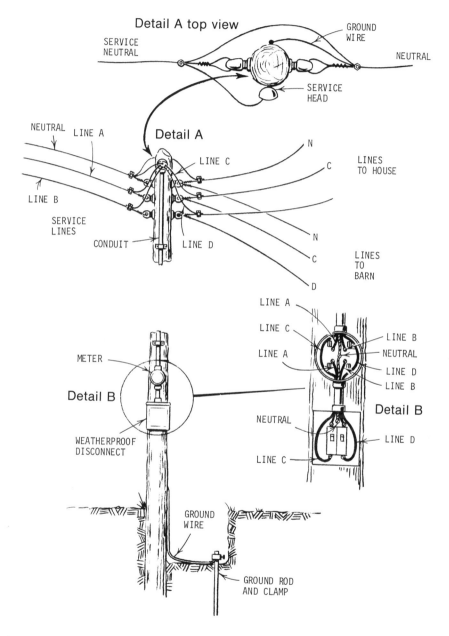

Of the two basic types of incoming and feeder line installations, this single stack (one conduit), once commonly used, is now prohibited by some localities.

SINGLE-STACK INSTALLATION

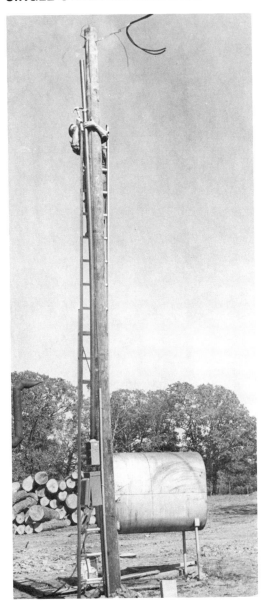

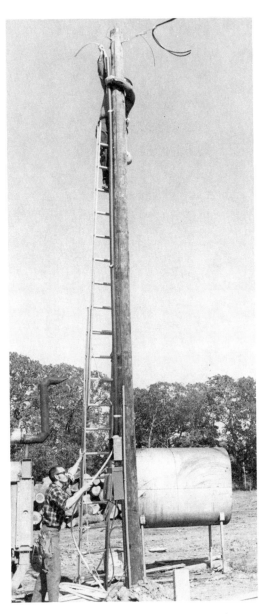

Here the meter housing and the disconnect box have already been installed. Feeder lines will enter underground. The bottom end of the conduit has been inserted into the meter housing, and the conduit's top bracket is being nailed into place. (Though materials shown here are heavy-duty—for a saw mill—procedures are the same as those for farms and ranches.)

With the conduit installed, feeding the service entrance cable through the conduit is a 2-man operation.

After fastening the base of the entrance head to the conduit and running the wires through, mount the top cover, as shown.

The Code requires that the top point of the service head be above the top insulator or other wire-holding device. *Caution:* Some local codes will not allow metered and unmetered wires to run in the same conduit. So make sure you check with local authorities.

The second installation is called a "double-stack," and in most cases it consists of either two conduits, or in most locales, two entrance cables with an entrance cable head on each end. In double-stack construction all three wires, including the neutral, are brought down through the conduit, or by an entrance cable of the proper size to the meter. Then all three wires are run up the second conduit, from the meter to the top of the pole. In this type of installation, the neutral wire from the incoming service wires is not spliced directly to the outgoing feeder neutral wires, but runs back to the top of

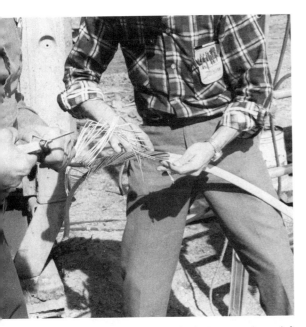

Strip the wires at the bottom, as shown left, and connect them into the meter housing and the separate disconnect.

DOUBLE-STACK INSTALLATION

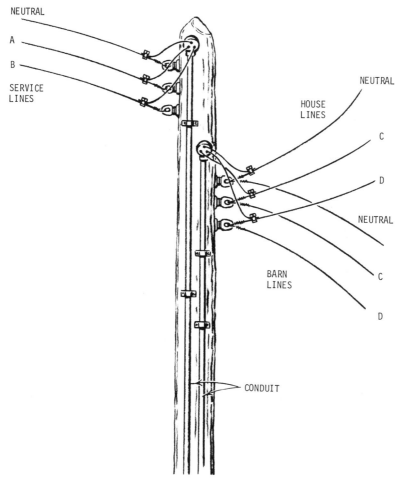

NEUTRAL

A

B

SERVICE
LINES

NEUTRAL

HOUSE
LINES

C

D

NEUTRAL

BARN
LINES

C

D

CONDUIT

If local code does not allow two runs of entrance wire in one conduit (one stack), use a double stack, as shown.

the pole and is then spliced into the feeder wires. The meter base is installed by driving screws into the pole, and the wires are attached in place, as illustrated in Chapter 8, on service entrances. From the bottom two terminals of the meter, the hot wires are run to the breaker switch installed on the pole.

One of the main things to remember in

pole hardware installation is to leave enough room so the lineman can easily climb the pole for future work. After installing the entrance head, service wire runs, and the switch, you can then call in the utility company to connect-in the power lines. By leaving the service switch off, you can connect the feeder lines. Or you can run all feeder lines before having the power connected in.

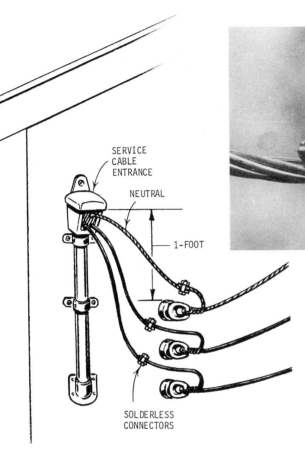

SERVICE
CABLE
ENTRANCE

NEUTRAL

1-FOOT

SOLDERLESS
CONNECTORS

Most local codes require that feeder wires connect to a building at least 1 foot below the entrance head, as shown in the drawing. This ensures that rain and melt water don't run into the head. Solderless connectors, shown in the photo, are required.

If the power is connected *make sure it is turned off at the pole breaker box before you install the feeder lines.* The weight of the wires, plus wind and the likely accumulation of ice in winter, dictates that the insulators be securely fastened to the pole. The wires should be tied around the insulators as shown in Chapter 8.

The yard pole must be grounded with a No. 6 or larger bare-copper grounding wire from the top, "neutral" wire of the incoming wires. The grounding conductor must be fastened to the pole every 6 inches by staples. In cases where the ground wire might be injured, run it through conduit as well. Some local codes require this.

The copper ground rod should be at least ½ inch in diameter and 8 feet long. Position it about 2 feet away from the pole and drive it into the ground with the top of the rod at least a foot below ground level. Clamp the copper grounding wire to the grounding rod using a solderless copper clamp. After installation of the yard pole, you're ready to run the feeder lines.

ABOVEGROUND FEEDER LINES

With the power shut off at the pole, connect the feeder lines to the separate buildings. Once again make sure you use the proper size wires and connect them securely to the pole insulators. Make each connection with a solderless copper wire clamp. Use only approved regular weatherproof wire. Normally No. 8 is considered adequate in most areas. If using individual wires, place the insulators on buildings at least 12 inches apart. All aboveground wires must be at least 18 feet above traffic areas such as roads, lanes and driveways. And they should be 10 feet away from trees and other obstructions.

At each building, install a regular entrance cable, head and all, and run wire into a service panel or fuse box. Some local codes may require a conduit run and service head instead of the entrance cable. Each building must also be properly grounded just as you grounded the yard pole, and it's a good idea to install circuit GFCIs in all damp locations. When grounding separate buildings, the grounding electrode must be connected directly to the neutral wire at the service entrance.

If local code allows you to run wire between buildings, be sure to follow directions carefully.

OVERHEAD CABLES BETWEEN BUILDINGS

In the case of small, closely clustered buildings, sometimes it's feasible to run overhead cables between buildings. In this case, install them in the same manner as you would install a service cable entrance on a house, as shown in Chapter 8. Or if codes permit, you can use the method shown in the photo, above right.

UNDERGROUND WIRING

The start of an underground feeder system on a farm is very similar to that of an aboveground system, with the metered pole carrying the incoming lines. The accompanying illustration shows a complete underground system with feeder lines running to a barn.

Today there are basically two types of underground wiring: 1) trench cable which is used for the larger wires normally used in underground feeder service to barns and 2) type UF underground feeder and/or branch circuit cable, in sizes 14 to 4/0.

The main thing to remember in underground wiring is that all bends in the cable must be enclosed in conduit where they're exposed, or in any location where there might be danger to them. Normally underground wires are buried about 24 inches deep. But on farms they're sometimes placed a bit deeper to prevent damage from the weight of heavy equipment. It's a good idea to dig the trench, and put in the wires between layers of sand. The wires should not be pulled tightly in the trench, but should be allowed to snake and wander a bit. This allows for expansion and contraction and prevents injury from surface stresses. The conduit carrying the cable underground from the circuit breaker on the pole should extend down into the trench. Its end should be fitted with a special underground conduit bushing. The cable is formed in an S loop before laying in the trench to prevent stress on the cable.

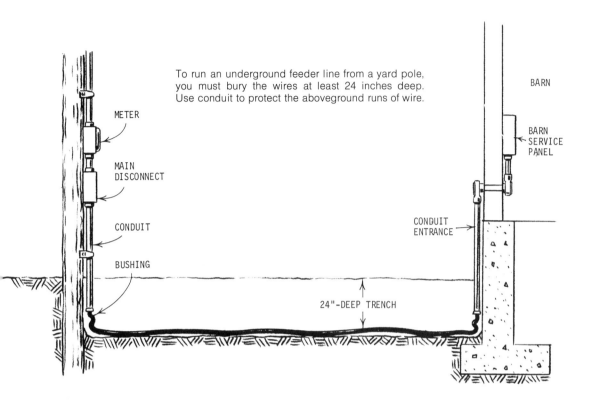

To run an underground feeder line from a yard pole, you must bury the wires at least 24 inches deep. Use conduit to protect the aboveground runs of wire.

METER

MAIN DISCONNECT

CONDUIT

BUSHING

BARN

BARN SERVICE PANEL

CONDUIT ENTRANCE

24"-DEEP TRENCH

Each separate building must have its own subpanel. Make sure you check with local authorities as to the types of materials and methods allowed for underground wiring.

FARM LIGHTS

The installation of rural lights is handled exactly as described in Chapter 15, on outdoor wiring, except that you should remember to include enough lights for areas where you will work after dark. Yard pole lights mounted at strategic points and controlled by 3-way switches in the house and barn are real lifesavers on the farm. Yard lights should be installed just outside workshops, at the entrances to all barns, and over areas where expensive equipment is stored.

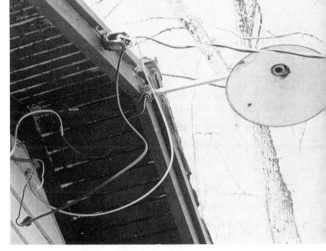

Outside lights are essential in rural areas. Placed at strategic places, they promote safety and security.

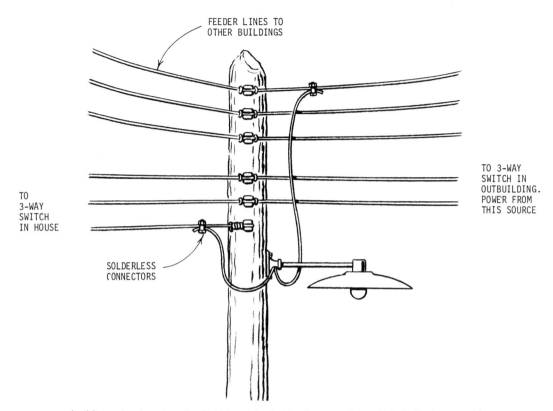

FEEDER LINES TO
OTHER BUILDINGS

TO 3-WAY
SWITCH IN
OUTBUILDING.
POWER FROM
THIS SOURCE

TO
3-WAY
SWITCH
IN HOUSE

SOLDERLESS
CONNECTORS

In this yard pole setup, the light is controlled by 3-way switches both in the barn and in the house. Note the manner in which the light wires are connected. Installation of 3-way switches is covered in Chapter 7.

BARNS

There are all kinds of barns, ranging from plain old general purpose barns to dairy barns and horse barns. Each type has specific wiring needs. The drawing on page 173 shows a floor plan for wiring a utility barn. But if you're interested in the complete wiring details for specific kinds of barns, contact your local county extension office for an applicable construction booklet.

TIPS FOR WIRING OUTBUILDINGS

Wiring a barn or outbuilding is basically like wiring a house, and it should be done just as carefully and with as good workmanship. Since in most cases the wiring in a barn will be exposed, care and correctness become especially important.

Just as in house wiring, run the cable so that it closely follows the contour of the building. This is done by drilling holes to run the wire through obstacles such as studs and joists, and by anchoring the wire to beams and joists. Follow code rules carefully. In dusty buildings such as hay barns and chicken houses, use lighting fixtures to

Here is an old-style fruit-jar bulb cover used in a dusty, dry hay barn. These jars were widely used to keep dust from igniting. Modern lights are available for this use and should be used in any area especially susceptible to fire.

minimize the entrance of dust, foreign matter, moisture, and corrosive material. Grounding is especially important in barns and other damp areas. So follow all good grounding rules carefully, as described in Chapter 8.

A great many fixtures in farm building wiring will be surface fixtures. Consult the illustrations on page 174 for proper installation procedures and circuit details.

WATER SUPPLIES

In most instances the pumps on a farm or ranch will be installed by the well driller or plumber. However, you must know how to repair them in the event of a breakdown. Or else you may have to install a new pump. The drawings on the following pages show the major types of pumps and the ways they

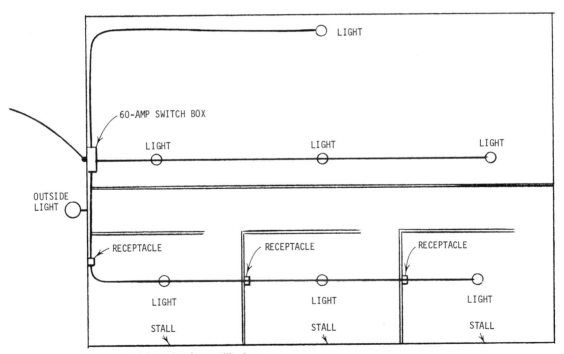

This is a lighting plan for a utility barn.

INSTALLING AN OUTBUILDING FIXTURE

In most outbuildings, surface-type switches, receptacles and fixtures are popular. To install a fixture, screw the base (without cover) into the building.

Using a pliers, break out the small knockout holes in the end and thread the wires through.

Connect the wires: White wire to silver terminal. Black wire to brass terminal. Ground wire to ground terminal.

After mounting the cover plate, thread an insulating ring over the metal bulb holder, as shown.

This chart by Leviton shows wiring schemes you may encounter in surface-mounted devices.

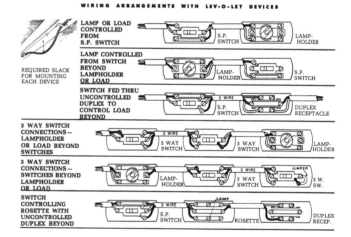

WIRING ARRANGEMENTS WITH LEV-O-LET DEVICES

REQUIRED SLACK FOR MOUNTING EACH DEVICE

LAMP OR LOAD CONTROLLED FROM S.P. SWITCH

LAMP CONTROLLED FROM SWITCH BEYOND LAMPHOLDER OR LOAD

SWITCH FED THRU UNCONTROLLED DUPLEX TO CONTROL LOAD BEYOND

3 WAY SWITCH CONNECTIONS — LAMPHOLDER OR LOAD BEYOND SWITCHES

3 WAY SWITCH CONNECTIONS — SWITCHES BEYOND LAMPHOLDER OR LOAD

SWITCH CONTROLLING ROSETTE WITH UNCONTROLLED DUPLEX BEYOND

For grounding No. 5238 in accordance with N.E.C. requirements, the grounding terminal (green) must be connected to a permanent ground. Such permanent grounds have been defined to include grounded outlet boxes, water pipes, conduits, metal frames of buildings and properly grounded driven rods and buried plates.

3-WIRE DUPLEX RECEPTACLES WITH 2-WIRE DEVICES.

3-WIRE DUPLEX RECEPTACLES ON SAME CIRCUIT.

note: When wiring for grounded receptacles, it is preferable to use a separate circuit.

are wired. You may choose to have the pump on a completely separate circuit from the house. Then in case of a house fire, the pump can operate.

LIGHTNING ARRESTERS

Lightning is one of the main dangers to rural wiring systems. This gives rise to a device called a lightning arrester. Normally lightning will not strike a wire directly, but even a close hit will allow the lightning's high voltage to follow the wires and damage or even destroy equipment like water pumps and television sets. Proper grounding will help alleviate the problem, but an inexpensive lightning arrester installed directly on the meter-pole service switch or on the service panel can help stop the problem entirely.

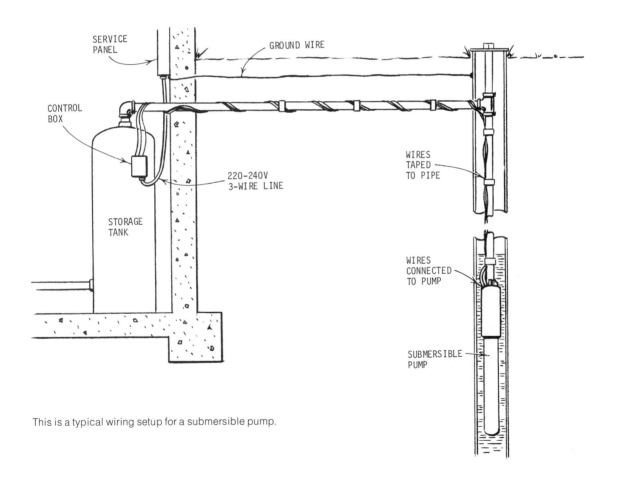

This is a typical wiring setup for a submersible pump.

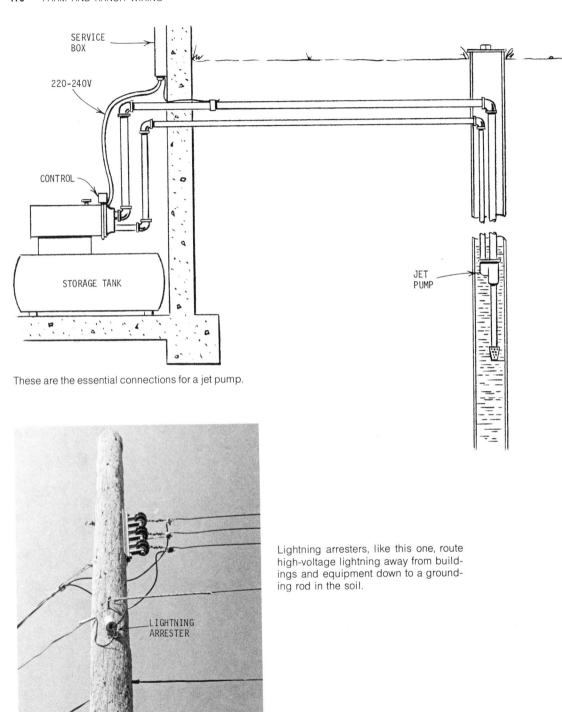

These are the essential connections for a jet pump.

Lightning arresters, like this one, route high-voltage lightning away from buildings and equipment down to a grounding rod in the soil.

14

Low-Voltage and Special Indoor Wiring

Today's electronic indoor gadgetry is utilized for everything from operating garage doors to piping music throughout the house. Although some of these systems appear complicated, installing them is actually quite straightforward. This holds true for door chimes, intercoms and even central vacuum systems. In fact, working on many of these household "gadgets" can be a great way of relaxing. Most of the work is fairly simple, requiring only the tools normally used for electrical wiring.

INTERCOM SYSTEMS

A good intercom system can be one of the most practical electronic helpmates in the home. It may consist of just two speakers run between the basement and kitchen, or it may include a system of speakers in each room for two-way conversation, as well as speakers for answering the front door and monitoring the baby. Many of the more complete systems also include such things as built-in radio, clock, tape deck and record player. Regardless of the number of units in a system, installation is basically the

same as that for a simple "squawk box."

The first step is to make a drawing of the house. Determine where you plan to install the central or master station, and where each remote station is to be located. Make a careful examination of the house construction if you're installing a system in an existing home, and plan for the most efficient use of materials, as well as for ease of installation. Check the walls for obstacles such as piping, heating ducts, and electrical power runs. The speaker wires should not be run through conduit already carrying electrical wires, or in channels and wire runs with other electrical wires, because the electrical wires cause interference and thus reception problems. So, you should allow at least 12 inches clearance between the speaker runs and any other electrical runs.

Whenever possible, locate the stations on inside walls rather than on outside walls. The condensation that often collects on the inside of outer walls can eventually ruin a unit.

A transformer is used to convert the 120-volt house power to 12 volts, and the transformer should be located at an easy access place. Ideally it should be located near the fuse or circuit breaker box in a utility room, garage or basement.

After deciding on the best areas for making the wire runs, determine the exact location of the master and remote stations. They should be located between studs so you will not have to cut through a stud and weaken the house construction. The master station should be located in a central portion of the house and in an easy-to-reach spot—usually somewhere in the kitchen. Normally it is positioned with about 60 inches from the floor to the top of the unit. This height is normally about right for remote stations as well. Never position stations back to back in a wall because their closeness will cause interference. And avoid locating outside remote stations in masonry walls.

Deluxe intercoms afford entire electronic control systems, including intercom controls and speakers in every room, AM-FM stereo radio, as well as tape and record players.

Modern homes often feature intercom systems that enable you to monitor children's rooms and talk to people in distant rooms, at the door, and even outside. Some systems also have built-in radios and clocks.

Estimate the amount of wire needed. The primary side of the transformer is wired with 14-2 wire to regular 120-volt AC housepower. The secondary side (wire to the master station) uses 18-gauge, 2-conductor wire for runs up to 50 feet. If the runs are over 200 feet, use 14-2 wire. Measure the total distance from each remote station to the master and add at least 36 inches to insure you have enough wire for proper connection to the units.

After planning all locations, and ordering your units and wiring, the next step is to cut out the openings for the master and remote stations. Using a stud finder, locate the studs. Mark the position of the stations so one side of the unit will be supported by a stud. Drill pilot holes at the corners of the rough-in frame opening of the unit and cut away the opening using either a hacksaw blade with one end taped, a sabersaw, or keyhole saw. Carefully remove the cutout so it will not break away from the wall. Install the rough-in frame. After cutting all frame holes and installing the rough-in frames, drill the wire-run holes into the ceiling plates and studs, and run the wires to the locations of the master and remote stations. This type of job involves the same problems as installing any other type of wiring in an existing home, and you'll often have to resort to some creative methods for fishing wires, as discussed in Chapter 10.

Access in some walls, particularly in two-story homes may be quite tedious. Wires to wall cutouts on the upper levels are run from the attic. Wires to wall openings on the first floor should be run from the basement or the crawl space.

One of the easiest methods of running wires into an attic is to locate the vent stack from the bathroom and run the wires alongside it or up the same wall.

Next, install the di-pole FM-AM antenna in the attic and run the antenna wire down to the master station as indicated in the instructions with your particular unit.

After you've run the wires and mounted the rough-in frames, install each unit and connect the wires according to the instructions with the unit. Connect the wires to the master unit and to the transformer, then *shut off the house power* and connect the transformer to the house power. Turn the power on again and check out the system to insure everything is working properly.

REMOTE STATION

1. After sketching a house plan, showing approximate locations of intercom units, you must select exact positions for rough-in frames. Since the frames must be nailed into wall studs, you'll have best luck with a magnetic stud finder. It points to nails and thus locates studs precisely.

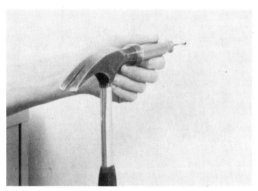

2. Punch-mark the edge of the stud on the side where the rough-in frame will be fastened.

3. Use the rough-in frame as a template and mark the wall for cutting.

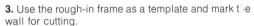

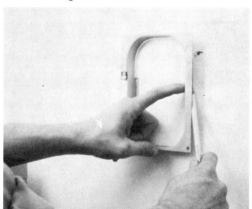

REMOTE STATION CONTINUED

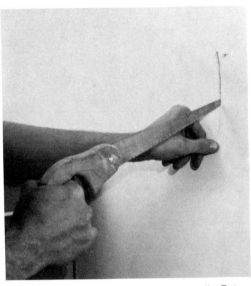

4. Use a keyhole or saber saw on most walls. But on plaster, a hacksaw blade with tape-wrap handle may prevent excess damage.

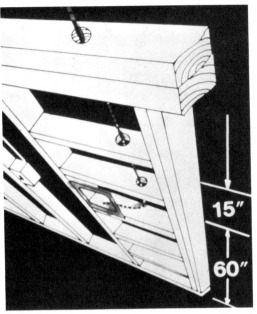

6. To install wires in an existing building, you'll need an extension bit to reach down from the attic and through fire stops.

5. Nail one side of the rough-in frame to the stud.

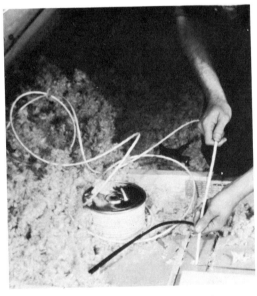

7. Feed wires down through the holes so they pass near the intended station.

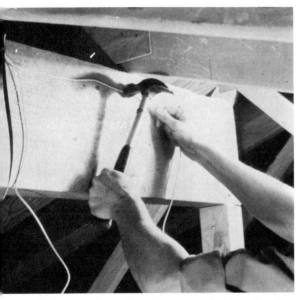

8. For a system with AM-FM radio, install a dî-pole antenna in the attic, tacking the antenna to rafters or support beams.

10. Connect the wires to the remote-station speaker terminals following manufacturer instructions.

9. Drop all remote station wires plus the antenna lead-in wire between studs to the master station.

11. Then install the speaker and finish panel over the rough-in frame, and one station is done.

PATIO STATION

1. For the patio speaker, drill holes at corner marks for your saw turns. Drill so that one side of the eventual opening will abut a stud.

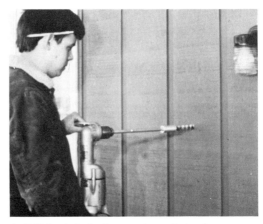

4. Drill the hole for the remote control unit.

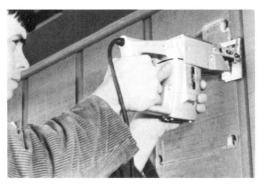

2. Use a saber or keyhole saw to cut the opening.

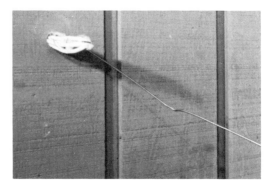

5. Using a bent coathanger at the remote-unit hole, fish out wire coming from the attic.

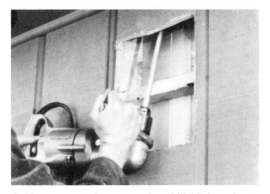

3. You may need an extension drill bit in order to reach into the attic.

6. Mount the control-unit shield housing.

7. Connect the wires to proper terminals.

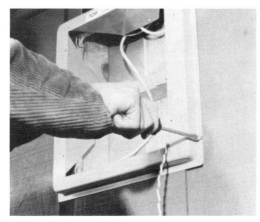

10. With the frame level, screw it to the wall panel.

8. Then mount the control panel.

11. Connect the wiring to speaker terminals.

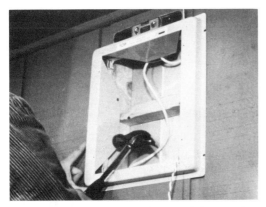

9. Place a small level at the top of the speaker rough-in frame and nail the frame to the wall stud.

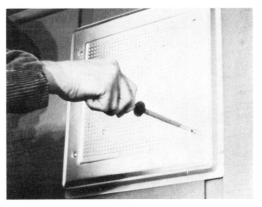

12. Screw-fasten the finish panel to the frame. (Use the same procedures for the front-door speaker.)

To get power for the system, you can mount a transformer to a junction box, say, in the garage. Then run wires from the transformer to the master station, normally located in the kitchen.

MASTER STATION

1. Install the master station just as you did the others.

2. Pull the wires down out of the wall from the attic, and screw the rough-in frame to the stud.

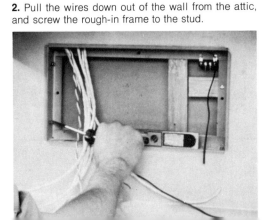

3. Screw the antenna lead wire to the terminal strip.

4. Screw the top mounting bracket to the frame.

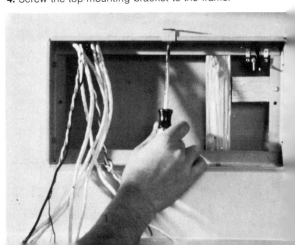

5. Mount the master unit to the frame.

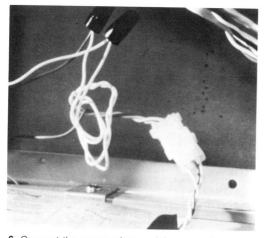

6. Connect the power wires and the ground.

7. Connect wires to proper terminals.

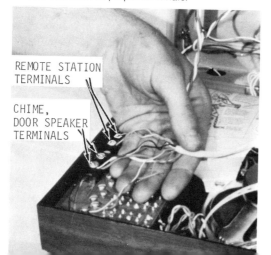

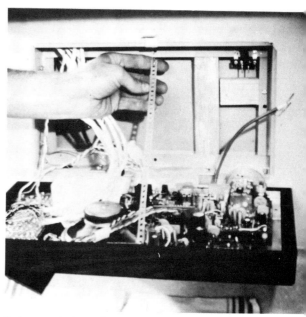

8. A support strap allows the unit to swing down for maintenance. Next step is to connect the antenna to the terminal strip.

9. Then close up the master control panel and proceed to test the system.

INTERCOM IN A NEW HOUSE

3. If stud spacing is too wide to secure the frame on both sides, install a filler block, as shown.

1. The best time to install an intercom is during house construction before the walls are in place. Here begin by leveling rough-in frames and fastening them to wall studs.

4. Nail the rough-in frame to the stud and filler block.

5. Connect the transformers to a junction box. Here the unit at the top of the photo is for a door chime and the other is for a radio-intercom system.

WIRE
PULLED
THROUGH
HOLE

UNIT
FRAME

2. After drilling holes through the top wall plates, pull the wires through.

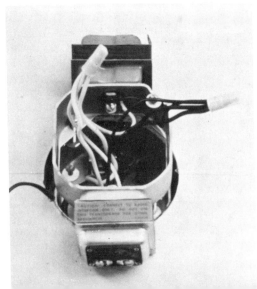

6. Place the master station so that its top measures about 5 feet from the floor.

9. Then wait to install the speakers until after the wall panels are in place.

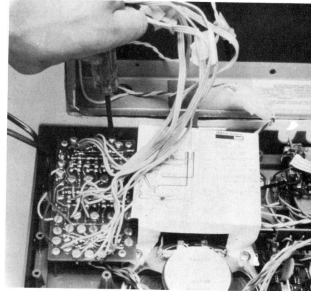

7. Using masking tape labels, mark each wire with the name of the room it serves.

8. Bring in wire from the transformer and tie it off in a junction box over the master panel. The junction box must be accessible.

10. Connect all wires to terminals in the master control station.

11. Secure the master panel in place and run a test.

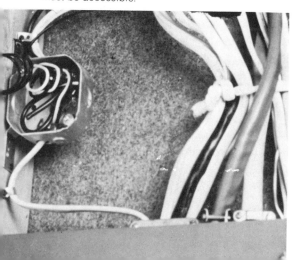

ELECTRONIC DOOR CHIMES

Today's electronic door chimes are a far cry from the old-time buzzers. For one thing, today's door chimes are beautifully constructed to add a decorative touch to almost any decor. Now you no longer have to hide an unsightly door bell or buzzer. Some of the more sophisticated units will sound two notes for the front door, one note for the back, and even a different signal for a third door. But most important, today's door chimes are easy to install.

It's very important to determine a good location for the chimes. Normally the unit should be placed at about eye level and in a central location so it can be heard from almost any location in the house. For large ranch type homes, split level homes or multistory homes, you may need to install two or more chimes in different locations.

Electronic door chimes like all low-voltage wiring must be installed according to local wiring codes. Check with your local electric utility company for a copy of the local code.

The first step is to locate and install the transformer that converts 120-volt AC house current to the low-voltage current needed for the chimes and door push buttons. *Before you begin make sure all house power is turned off.* The easiest means of installing the transformer is merely to attach it to a junction box. Most transformers are fitted with an angle bracket that has a screw in it. Remove an appropriate knockout hole from the junction box. Insert the bracket and wires in the knockout hole, and tighten the screw on the transformer to secure the transformer to the box.

Run 120-volt AC current into the junction box and (if your local code allows) use

Deluxe facades for electric door chimes are often mistaken for nonfunctional decorator items.

twist-on connectors to connect the transformer wires to the house current wires. Run the wires from the transformer to the chimes location, as well as to the door-bell location. *Caution: Do not connect to the transformer yet.* Turn the house current back on.

Use 2-conductor 18-gauge wire from the transformer to the house current wires. Professional installers use furnace thermostat wire, because it is the easiest to install. But make sure you check with local codes as to wire requirements and transformer/junction

INSTALLING DOOR CHIMES

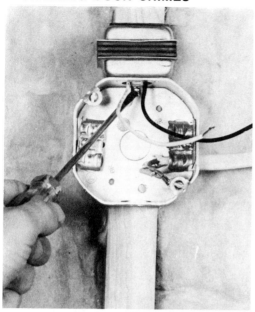

1. *First, turn off house current.* Then install a transformer on a junction box near the unit and run-in 120-volt house current.

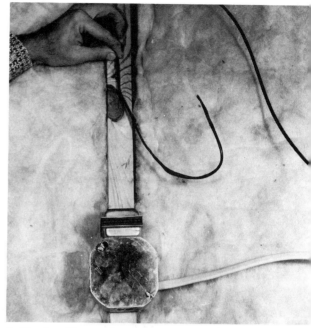

3. After installing the box cover plate, run wire (without hooking it up) to chime and push-button locations.

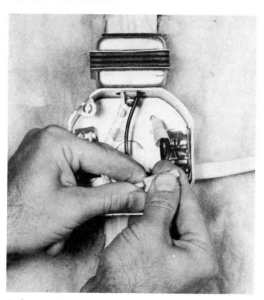

2. Connect house current wires and transformer wires, black-to-black and white-to-white.

4. Hold the chime unit in place and mark the wall for the wire entrance hole.

5. After boring through the wall, fish the wire out by means of a loop-ended coathanger.

7. Connect wire to the terminal board according to the manufacturer's directions.

6. Mark and then drill for the unit's top holding screw. After the screw is in place, hang the unit on it.

8. Then merely slip the cover over the unit.

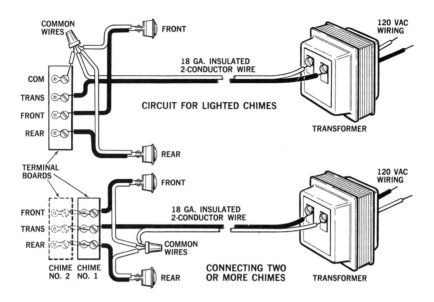

COMMON WIRES

FRONT

120 VAC WIRING

COM

TRANS

FRONT

REAR

18 GA. INSULATED 2-CONDUCTOR WIRE

CIRCUIT FOR LIGHTED CHIMES

TRANSFORMER

REAR

TERMINAL BOARDS

FRONT

TRANS

REAR

CHIME NO. 2

CHIME NO. 1

FRONT

18 GA. INSULATED 2-CONDUCTOR WIRE

120 VAC WIRING

COMMON WIRES

CONNECTING TWO OR MORE CHIMES

REAR

TRANSFORMER

9. At each push-button location, drill a ⅝-inch hole. Then fish the wires through and connect them to the terminals. Press the push button into the hole.

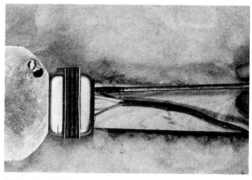

10. Last, with house current still off, connect the wires to the transformer terminals.

box locations. Most codes don't recommend an attic location. So you may have to install the transformer in the garage or basement, or near the circuit breaker box.

When installing the wires in a new unfinished home, you can drill holes down through the plates at the top of the walls and drop the low-voltage wire down between the studs in the approximate location of the chimes or push buttons. For this reason, it's much easier to install the chimes on an inside wall. In an older home the installation of the chimes and door-bell wire is a bit harder. If the house or walls will receive new plaster, wallpaper, or sheetrock, you can chisel out channels to drop the low-voltage wires into the walls from the attic. Then cover the wires when applying the wall cover. Staple the low-voltage wiring to ceiling joists or wall studs, making sure that you don't drive a staple through or cut the wiring, which might cause a short.

If you have an older chime system that merely needs replacing you've got it made. Tape the new wires to the old and pull the new wires into place as you pull out the older wires. When replacing a unit, always replace the wires of the older unit with those of the new one.

With the low-voltage wires run to the intended locations, install the chime unit according to the manufacturer's instructions. Most chime units come with an inside shell, containing the chime mechanism, and an outside shell which is installed after wiring is completed. After running all wires, you can install these units over a new or finished wall with little fuss or mess.

Hold the inside unit in position on the wall and mark for the wire entrance hole. Drill a ⅝-inch hole through the wall with a paddle or spade bit. Using a piece of coathanger wire with a small hook bent on the end, fish the low-voltage wire out through the hole. This normally takes some patience and maneuvering, but don't give up. Once the wire is pulled through, run it through the entrance hole in the unit, and mark the position of the top holding screw on the unit. Bore a small starting hole for the screw and fasten the top of the unit in place. Lay a small level across the top of the unit, and when the unit is level bore a small starting hole for the bottom screw. Fasten the unit securely to the wall. Connect the wires to the terminal board of the chime according to the manufacturer's instructions.

Make sure the wires do not touch the tone bars or the power unit of the chime. With the wires installed in the chime, slip the chime cover or outside shell over the base plate or inside unit. For surface-mounted, lighted chimes, connect the common wires from the transformer and push buttons to the common terminal on the terminal board. (Note: When using lighted chimes make sure you use a heavier transformer that is specially suited for that purpose.)

Installing the door-bell push buttons in an older or finished home can be a bit of a problem. The hard part is drilling the holes in the top plate of the outside wall to drop the low-voltage wires down to the push button location. You may have to buy or rent an angle adapter for your drill to give you enough clearance to drill. Naturally in new homes this can all be done before the walls are finished.

Once the holes are drilled and the wires dropped down, the hard part is over. Drill a ⅝-inch hole in the location for the push button and once again using a coathanger wire, fish out the low-voltage wire. Attach it to the terminal screws on the push button and gently press the push button into the hole. If you prefer a fancy escutcheon plate, install it now.

With all wiring run and connected to the chime or chimes and push buttons, *cut off the house power and connect the low-voltage wires to the transformer.* Turn on the house current again and check the unit. If the chimes do not operate, turn off power and check to make sure you have wired it according to the manufacturer's instructions. If the circuits are wired correctly and the unit still doesn't operate check the push buttons for poor contact or loose connections.

To keep your new electronic door chime operating properly, occasionally clean it with a nonflammable cleaning fluid, and then wipe dry.

DUCTED HOOD VENTILATOR

Today's modern kitchen normally has a system for venting odors and dangerous fumes outdoors. These ventilating units are easy to install, particularly in new construction. Again in old or existing construction, it takes a bit more planning and carpentry to install the unit without damaging the structure of the building.

Cut the openings for the vent ducting as described by the manufacturer and knock out the proper holes in the hood to mate

A ducted ventilator hood over a stove can remove unwanted cooking odors and airborne grease.

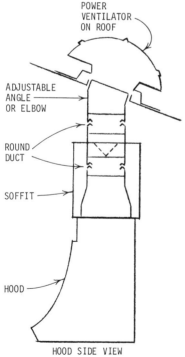

HOOD SIDE VIEW

These are hood-ventilator components you would be likely to encounter.

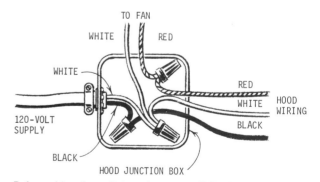

Before wiring the unit, *turn the power off.* Run house wiring into the junction box and secure it with a standard clamp. Connect house and unit wires— black-to-black and white-to-white. Connect ground wires in accord with local code. Then mount the junction-box cover and turn on the house power.

with the ducting. Install the ducting and mount the hood in position. The hood should be installed about 24 inches from the stove or range top.

Next, *turn off house power.* Remove the cover from the junction box on the unit, and bring 110–120-volt AC house power with ground into the hood junction box. Allow at least 6 inches of wire to protrude from the box for ease of connecting, according to the manufacturer's instructions.

Replace the junction box cover and turn on the house power again.

CENTRAL VACUUMING

As with other systems described in this chapter, carpentry is the main job in the installation of central vacuuming. If the installation is done during the building of a new house, the job is fairly simple. However, installation in an existing house can be a headache. Once the mechanical portion of the system is in, wiring the system is quite simple. Most of the central power units are

A central vacuum cleaning system can eliminate much of the lugging, tugging, and bruising normally involved in keeping floors and furniture tidy.

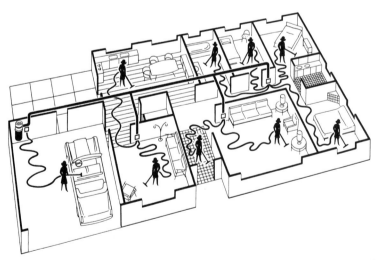

This illustration shows the relation of the vacuum-power unit to room inlets and runs of plastic tubing.

The central power unit should be installed in an out-of-the-way location that provides easy access for removal of the dirt and lint bag.

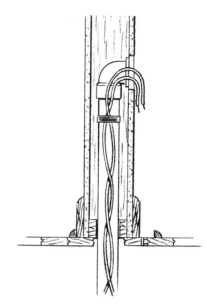

In an unfinished house, the first task is to install wall inlet rough-ins. Then run tubing from the power unit. Tape low-voltage wiring along the outside of the tubing as you go. The wires allow you to activate the power unit from any room.

equipped with a 6-foot grounded cord. The power units are normally installed in a basement or utility room and simply plugged into a 120-volt grounded receptacle.

First, you run a 2-conductor low-voltage wire to the terminal screws on the back of each automatic wall inlet. At the same time, you can be installing the tubing for each inlet, taping the wiring to the tubing as you go. The low-voltage wires are then connected to the terminal board outside the power unit.

ELECTRONIC GARAGE DOOR OPENERS

Again, most of these units merely plug into a grounded 120-volt house power receptacle. The major portion of the job is the installation of the mechanical parts.

Some automatic garage-door openers plug into an existing receptacle, as shown. Others are wired directly into junction boxes.

15

Outdoor Wiring

Outdoor lighting has changed a great deal in the past few years. At one time it was used only to provide light at dangerous walks, steps, and to ward off prowlers. Today, outdoor lighting, as well as outdoor wiring is much more versatile, and with a little planning it can make your entire backyard a giant work and play area.

Waterproof outdoor receptacles can provide electricity to run televisions, hi-fi equipment, barbeque grills, even electric lawnmowers and hedge clippers. Outdoor lights have changed as well. In addition to providing protection, they can light a play area such as a basketball court near the garage, a patio area or a pool. And low-voltage lights can even be used to accent fountains, pools, shrubbery and statuary.

Outdoor wiring is of two separate kinds: normal 120-volt house wiring and 12-volt, low-voltage wiring. Here are several safety rules that must be followed during installation:

1. According to the Code, all 120-volt, single-phase 15- and 20-amp receptacle outlets installed outdoors must be protected by a ground-fault circuit interrupter. This includes underwater swimming pool lights operating at over 15 volts.

2. Always check local code rules pertaining to any outdoor installation, and follow the rules carefully.

3. Always use the cable required by your local codes: either Type UF lead-covered, or Type NMC dual-purpose indoor-outdoor plastic-covered cables. If using lead-covered cable, make sure it is properly grounded.

4. In some areas you may be required to make the complete installation using conduit. Even if this is not required by local codes, you should use conduit at any locations where the cable is exposed above ground.

5. Make sure you bury the cable below the frost line or at least 24 inches beneath the surface to protect it from the accidental cutting from tools or the weight of heavy equipment.

In addition to their security and safety values, outdoor lights increase recreational opportunities. And properly wired outdoor receptacles can handle the gamut of electrical devices available today.

6. Use only approved waterproof receptacles, lights and boxes for all outdoor wiring.
7. *Make sure you shut off the current before starting any wiring project.*

OUTDOOR RECEPTACLES

Several good waterproof receptacles placed at strategic places around the house and yard can make any outdoor chore much easier. These handy receptacles can be mounted directly on the outside wall of the house or attached to stakes, rock walls, or outbuildings.

It's a simple matter to install an outdoor receptacle directly to the house. Merely cut a hole in the wall (outside sheathing and siding) making sure the opening does not hit a stud, and that there is easy access from the house power to the opening. Run the wire from the outlet opening to the service panel, but *do not connect it in*. Install a *waterproof* box; then connect the wires to a receptacle. Weatherproof covers are made in several different ways. Most of them have a rubber gasket that fits over a metal cover. The metal cover is fitted with plastic covers that snap over the receptacle when it is not in use. Finally, connect the receptacles to a circuit protected by a ground-fault circuit interrupter (GFCI).

Once again it is very important that GFCIs be installed on outdoor receptacles. They're somewhat expensive, but more than worth it for the protection they provide.

INSTALLING AN OUTDOOR RECEPTACLE

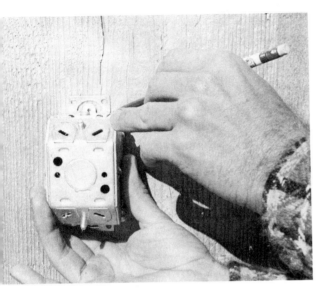

1. To begin installing an outdoor receptacle on the side of a building, use the box face as a template.

3. After connecting the ground wire, install a foam-rubber gasket. Then mount a special outdoor-receptacle box cover over the gasket.

2. Install the box and receptacle. Note: The box shown is not weatherproof. Local codes vary as to box-weatherproof requirements for buildings.

4. Here the receptacle covers are open. Note the foam gaskets within the covers.

5. When the box is not in use, simply snap the covers back over the receptacle.

GROUND-FAULT CIRCUIT INTERRUPTERS

Now required on all outdoor receptacles, a GFCI can detect and stop an unwanted current flow (ground-fault) between an ungrounded conductor and a ground. In other words, the GFCI stops the dangerous flow of current that may be diverted from the neutral wire, through a person or object touching the receptacle or appliance, and then to a ground. Note: A 50 milliampere shock can kill a healthy adult, and a "Class A" GFCI can shut off flow within .025 second of the onset of a leak as small as about 5 milliamperes.

Basically there are three different types of ground-fault circuit interrupters. One simply plugs into a standard 120-volt, 15-amp receptacle. Another is a GFCI receptacle which is used to replace a standard receptacle. The third is a circuit breaker GFCI which acts to protect the full circuit on which it is installed, and it may be permitted to protect a panel feeder circuit. Most of the better units may be tested by pushing a button to simulate a ground fault. Resetting another button reactivates the unit.

GFCIs are not intended to replace circuit breakers or fuses. These units are designed to protect a circuit's overloading and overheating while the GFCI device is a means of protection from a line-to-ground shock.

GFCI Installation

To install an outdoor receptacle at a location other than on a house wall, you'll have to run an underground cable carrying the house current to the location of the receptacle. Again check with local authorities as to what types of materials are required for underground wiring. Dig a narrow but deep trench from the position of the house current to the location of the receptacle. Drill a hole through the house siding and install a conduit and waterproof conduit ell. Then install a second piece of conduit on the underside of the ell. The bottom conduit should have a bend to fit at the bottom of the trench. Connect the wiring into the fuse box and thread it through the upper conduit, into the ell, and down through the lower conduit. Lay the cable into the trench. The conduit should be fastened to the house foundation by conduit straps. At the end of the run re-

INSTALLING A GFCI OUTSIDE

1. Begin by removing the old receptacle.

2. On vertical boxes, you'll have to install a GFCI support plate with adapter like this one.

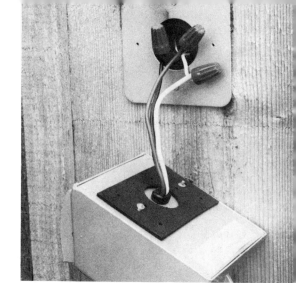

3. Using twist-on connectors, connect the GFCI to the circuit wires—white-to-white, black-to-black, and ground-to-ground. Mount a weatherproof sealer plate on the GFCI's back panel.

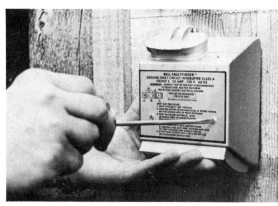

4. Stuff the wire and connectors into the wall opening, and screw the GFCI onto the wall support plate.

5. Once a month, test the unit by pushing the test button. This creates a ground fault that should cause the unit's circuit breaker to switch off. Then merely switch the breaker on again. On this unit, the receptacles are on the underside.

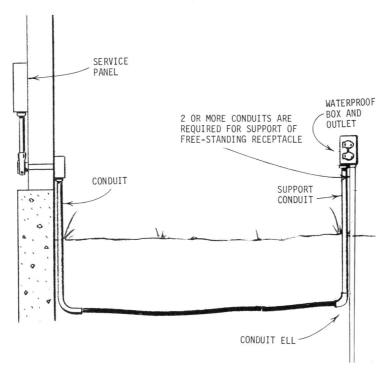

SERVICE PANEL

2 OR MORE CONDUITS ARE REQUIRED FOR SUPPORT OF FREE-STANDING RECEPTACLE

WATERPROOF BOX AND OUTLET

CONDUIT

SUPPORT CONDUIT

CONDUIT ELL

Like the receptacle shown, outdoor receptacles that are not on permanent buildings must be installed with dual-purpose cable. For ease of installation, some codes allow the "T" connection shown. Others may allow only conduit bends. Here it's a good idea to install a GFCI on the circuit inside the house or at the receptacle.

sulting in the outlet, use another piece of conduit with a bend and a support conduit. On the top of the conduits, install a weatherproof box, then a receptacle with a weatherproof cover, and fill in the trench.

Then *shut off the house current* and connect the circuit to the house current. In some cases local codes may require that the entire run be of conduit. To install a similar unit on an unattached building use the same methods, except the receptacle could be mounted outside the building, instead of inside.

POST-LANTERN YARD LIGHT

You can install a front yard post lantern in much the same manner as a receptacle.

Run a conduit or weatherproof plastic-sheathed cable to the location of the post. Dig a hole about 2 feet deep using an ordinary posthole digger. Dig the trench and run the wiring, leaving plenty of wire to extend up into the post to connect to the light fixture. At the location of the post pipe, place a bend of conduit and run the wire up through it. Place the post in position in the hole down over the conduit bend, bringing the wire through the post to the fixture, and pour concrete around the post. Tie or brace the post in position until the concrete sets permanently. Make sure the post sets plumb in all directions. After the concrete has set, strip the wire and connect it to the light fixture. Then install the fixture on the post.

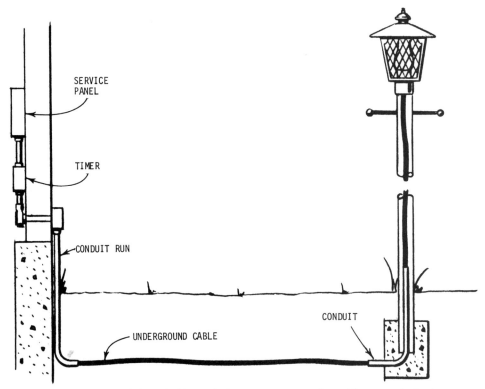

A post yard lantern is installed in much the same manner as the outdoor receptacle shown in the previous drawing. Here you may want to employ a wall switch, a timer, or else a photoelectric eye that can switch the lantern on and off at dusk or dawn.

Fill in the trench. After connecting the light fixture to branch-circuit conductors, *turn off the house current* and connect the branch-circuit conductors to the circuit breaker or fuse, or to an existing branch circuit. Then turn the power back on.

In some instances you may wish to install an automatic switch to turn the post light on at dusk and off at dawn. You can either use a permanent time switch installed in the house near the fuse box, or you can install a simple light-activated, photocell switch in the lamp socket of the light fixture or on the lamp post.

The main factor in installing such a light, as well as any outdoor light, is that the entire fixture must be grounded, including the (metal) post or anything else that may be touched accidentally. If you use conduit, some local codes may consider it the ground to a grounded box. If you use cable, *make sure the cable has a ground wire.* The ground wire must be properly bonded to the post.

Similarly installed lights may be used as yard or play lights, as well as security lights. Lights such as security spotlights in protected areas under eaves may be weatherproof fixtures. These are installed in much the same

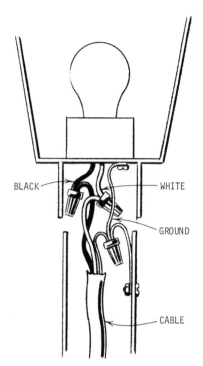

BLACK

WHITE

GROUND

CABLE

A post lantern is wired as you would wire any other light fixture, except that all portions of the post must be grounded by bonding the ground wire as shown.

Here are the timer switch (top) and the switch activated by photocell.

manner as indoor light fixtures; however, the box should be weatherproof. Preferably, security lights should be in high locations so they throw a wide pattern of light effectively. It's a good idea to connect a security light with two 3-way switches, one inside the house and one outside or in the garage. By using the combination, you can turn on the light from an unattached garage, walk safely to the house, and then turn the light off. Another good idea is to install a seven-day, 24-hour time switch on the security lights. The time switch can be used to activate lights while you're away.

INSTALLING A PORCH CEILING LIGHT

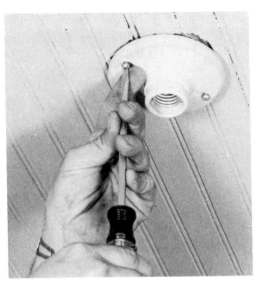

1. First, disconnect the old fixture from the ceiling, as shown. Then disconnect the appropriate fuse or circuit breaker at the service entrance panel.

2. Use a tester to check for hot wires. Then disconnect the wiring from the old fixture, as shown.

3. Install the hanger bracket for the new light.

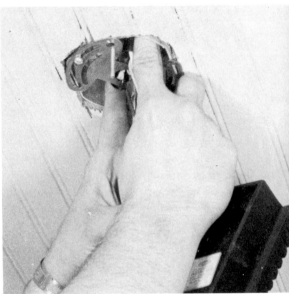

4. Connect the wires and then gently push them up into the box.

5. Fasten the new fixture base to the hanger bracket.

6. After installing the bulb, mount the cover.

INSTALLING A PORCH WALL LIGHT

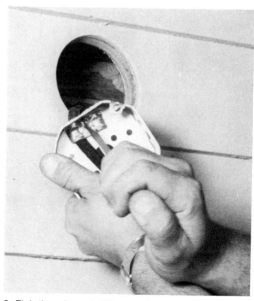

1. The best tool for making the opening for the box is a large "hole saw," making sure the fixture will adequately cover the hole and box assembly.

2. Fish the wire out of the opening and pull it through the metal box. In some cases, it's better to install a weatherproof box.

3. Fasten the box against a wall stud.

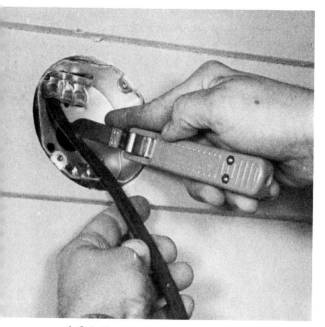

4. Strip the wire.

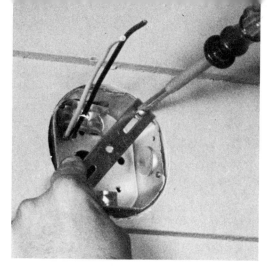

5. Mount the hanger strap.

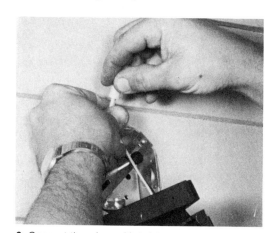

6. Connect the wires with twist-on connectors.

7. Fasten the top portion of the fixture to the strap.

8. Install the glass inserts in the bottom portion. Then screw in the bulb.

9. Fasten the upper and lower portions. (For details on installing switches, see Chapter 7.)

VAPOR LIGHTS

One of the newest forms of security lights is the mercury vapor light. You see this type most often in rural areas. These are connected directly to the power line coming from the utility company—the light merely being rented from the company. Mercury vapor lights are also available from electrical supply companies, as well as larger hardware stores. And they're great for lighting a large area such as a tennis court, swimming pool, or farm lot. Most are equipped with a photoelectric cell which turns them on at dusk and off at dawn.

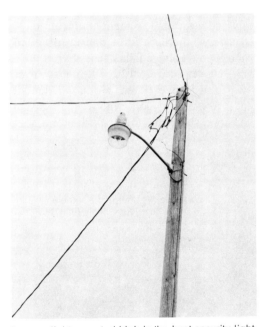

A vapor light mounted high is the best security light for large areas. It's also excellent for recreation areas.

LOW-VOLTAGE LIGHTING

If you wish to dramatize statuary, a patio or even foliage, a portable low-voltage lighting system is the answer. These little lights come in a variety of shapes from soft-glow types to little spots that add just the right touch to your favorite plant or statue. Low-voltage lighting is often called "low-key" lighting because the lights are somewhat dim and have a glowing effect that glamorizes your garden. Make sure you light any hazards in the garden, such as steps as well as the landscaping points you wish to show off.

A most important feature of low-voltage lighting is safety. If the unit is installed properly and according to the manufacturer's instructions, there is little danger of a bad shock. Low-voltage lights actually run on 12 volts converted from 120-volt house current by a transformer. Even if you accidentally cut a low-voltage cord with a spade or garden fork, there may be a spark and a slight

A patio that looks good by day has potential for glamorous night use. The keys are planning locations and selecting from lighting options.

With the weatherproof, outdoor switch components, below left, the lever on the cover switches the toggle inside the gasketed box. The indoor switch, at right, features an on/off indicator light.

This low-volt system from Montgomery Ward includes a transformer/timer unit, wire, and sealed-beam stake-mount lights.

This 12-volt assembly features a drive-in receptacle that will accept a variety of light posts.

The socket photocell from Sears can be inserted between a regular light and socket. The adjustable cell will turn the light on at dusk, off at dawn.

This is a common way of running low-voltage outdoor lights. Note the components.

SWITCH

HOUSE CURRENT

WEATHERPROOF
RECEPTACLE

MUSHROOM
LIGHT

SPOTLIGHT

TRANSFORMER

12V LOW-VOLTAGE WIRE

tingle but no dangerous shock. The lights are run from a transformer that is connected to the house current. And again, it's a good idea to connect a ground-fault interrupter between the house current and the transformer to provide added protection around the transformer. A normal size transformer would be 100 watts. This is large enough to run from 6 to 10 lights, depending on the size and the amount of light you wish. Install a lighting control switch inside the

Place the transformer near an outdoor weatherproof receptacle, but where it won't be accidentally touched by anyone.

Connect wires from the house to a transformer which converts current from 125 volt to 12 volt.

Connect individual low-voltage lights to the wire and simply push them into the soil.

From there it's a matter of stringing low-voltage wiring up trees, along trenches beside walks and patios, then to the fixtures.

The light fixtures are made in many different shapes and styles, and may be positioned in trees to throw light down onto the patio. Mushroom-shaped lights may be used as subtle lighting for sidewalks and patios, and as spots for highlighting outdoor decor. Garden-type lights are portable and are equipped with metal or plastic stakes which you can merely push into the ground in the appropriate place. Position the lights where you desire them, and connect the wiring to the fixture terminals.

You can lay low-voltage wiring in shallow trenches alongside sidewalks, bury it in concrete (if run through conduit), or place it in a bed of sand under a brick or flagstone patio. Lay the low-voltage wires in trenches at least 6 inches deep if possible. The simplest method of running a low-voltage wire across a lawn is to simply cut a slit with a spade, lay the cord in place and tamp the turf back in place. To connect the wires to a tree, use small staples. Or if you're afraid of damaging the tree, tie the cord in place with fine copper wire.

The lessened shock danger of low-voltage lines allows you to implace them with fewer code restrictions. You may want to run them under patio bricks and push them into slit cuts in the lawn. Or you can string them in trees.

house on the house current, in line ahead of the connection to the transformer.

Low-voltage lighting is extremely easy to install. Some transformers merely plug into a weatherproof outdoor receptacle; others are wired directly into the house current. In any case the transformer should be securely fastened in an out-of-the-way place such as behind the house shrubbery or in a corner. And it should be located for easy access to house current.

Install the transformer first, but do not hook it up to the house current. Then connect the low-voltage wires to the transformer according to the manufacturer's instructions.

FOUNTAINS

Fountains are another attractive outdoor feature. They can be run by a small low-voltage pump and can be lighted by low-voltage wiring as well. Again, as in the case of the low-voltage patio lights, install the circulating pump and lights. Then connect to the transformer, and finally connect the transformer to the house current. Although the low-voltage lights and fountain equipment take only a bit of electricity to operate, they're a joy to behold and can make even the most mundane back yard look exotic at night.

You may want to combine the effects of light and water in a garden pool or in a fountain, as shown here. Note: A GFCI should be installed on the circuit.

In any application concerning water and electricity, again make sure you know what the local code states regarding materials to be used, and comply fully. You should also install a ground-fault circuit interrupter in the circuit in accord with the codes.

SWIMMING POOLS

More and more people are turning to their back yards for play instead of taking vacations, and a good swimming pool can make a back yard even more fun and inviting. But it can also be a death trap, unless all electrical wiring is very carefully done, and the proper procedures are followed exactly. Because of the inherent danger, the Code is very strict about swimming pool construction. Check with local authorities as to what is required in your area for the various electrical components of the swimming pool. Naturally, all aboveground lights anywhere in the vicinity of the pool should be low-voltage or protected by a ground-fault cir-

cuit interrupter. All lights and switches should be located so there is no danger of a wet person contacting them. Receptacles should be at least 10 feet from the inside walls of a pool and must be protected by a ground-fault circuit interrupter.

The installation of wiring for a portable aboveground pool is actually quite simple. The portable filter pump normally plugs into a grounded weatherproof outdoor circuit which must be protected by a ground-fault circuit interrupter.

Permanent underground pools, on the other hand, have a much more complex filter system and sometimes have underwater lighting and an audio system, as well. Installation of these should be left to experts. If you contemplate this type of job, study the Code carefully before deciding to do the job. Otherwise, turn the job over to a professional electrician who specializes in swimming pool wiring.

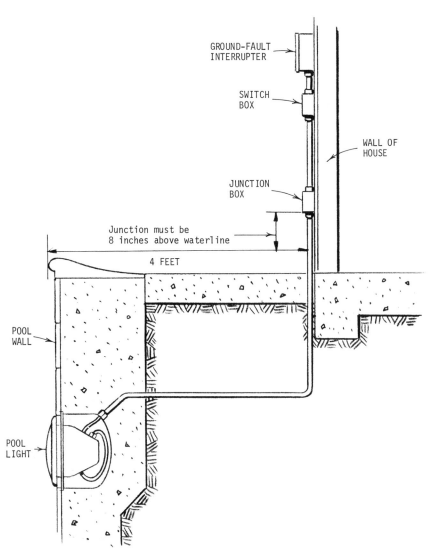

GROUND-FAULT
INTERRUPTER

SWITCH
BOX

WALL OF
HOUSE

JUNCTION
BOX

Junction must be
8 inches above waterline

4 FEET

POOL
WALL

POOL
LIGHT

This is a typical setup for underground lighting of a swimming pool.

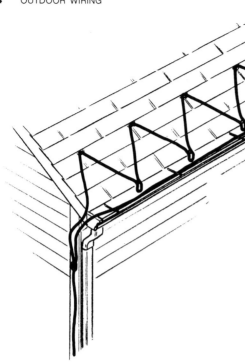

HEAT TAPES

One additional outdoor electrical item that can be a real headache solver is heating tape. This versatile electrical device can be used for everything from heating sidewalks to heating roofs and drains, thereby preventing ice and snow buildups. Heat tape can also be used in mobile homes to prevent water pipe freeze-ups, as well as in garden hotbeds. Each application requires a specific type of tape. So check with your local hardware store or electrical supply dealer as to the best type of tape to use. And again always make sure the tape is plugged into a properly grounded and GFCI-protected outdoor outlet.

Heat tapes have a variety of applications. They are commonly used for de-icing roofs and for heating garden hotbeds.

SECTION CHART for heating cable, tapes and thermostats

APPLICATION	LEAD-ARMORED CABLE		VINYL-JACKETED CABLE		HEATING TAPES		ROOF TAPES	THERMOSTAT CONTROLS
	Plug-in	Ready-to-wire	Plug-in	Ready-to-wire	Auto	Non-Auto		
PIPE HEATING*	X	X	X	X	(X)	X		X
ROOF & GUTTER Commercial Residential	(X) (X)	(X) X	 X				X (X)	
SOIL HEATING* Large Areas Cold Frames	X (X)	(X) X	X X	(X) X				X X
WATER TROUGHS*	(X)		(X)					X
SNOW MELTING		(X)		X				
POULTRY & PIG BROODING*		(X)		(X)				X

NOTES: (X) = Best for application.   X = Can be used.   * = Thermostat control desirable. (See far right column.)

(Chart by General Electric.)

16

Testing and Troubleshooting

There are three kinds of electrical wire testing. You can 1) test to determine if a circuit is hot before you work on it, 2) test a new circuit to see if it is working properly, and 3) test or troubleshoot to find a specific problem in a new or old circuit.

BEFORE WORKING ON A CIRCUIT

Before working on a circuit, test it to be sure that it is "dead." This precaution should be taken even if you have shut off the circuit by flipping a circuit breaker or removing a fuse, since sometimes the labeling of circuits on a service entrance panel is mixed up. Many people shut off a circuit and confidently start to work on it, only to get a bad shock. Even if you feel certain that a circuit is shut off at the service panel, use one of the small 2-prong light testers to make sure. If you're working on a receptacle, unscrew it from the box; gently pull it out of the box *without touching the terminals;* and touch one prong to each of the terminals. If the tester lights up, the circuit is still hot. If the tester doesn't light up, the circuit is dead.

The warning light on this 2-prong tester could save your life. Always check to be sure a circuit is "cold" before working on it, even if you feel certain you shut it off at the service panel.

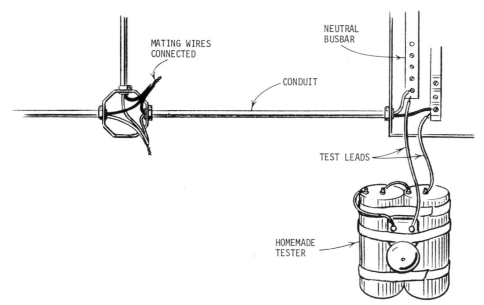

This homemade device allows you to make the several tests on unenergized circuits described in the text.

TESTING A NEWLY INSTALLED CIRCUIT

There are two methods and several different tools for testing a circuit. The first method involves testing the circuit before it is completed. In other words, the wires have already been run in place but have not been connected. *Note: The incoming current from the service panel must not be turned on by circuit breakers or fuses.* This type of testing is especially important when you are running conduit or armored cable wiring and you have a number of different colored wires.

For uncompleted circuits, you can make a tester from 2 dry-cell batteries and a door bell. Run a couple of test leads from the terminals. Before you make the test go around to each outlet and junction box and

gently twist or loop the wires together, as they would normally be connected together. Do not connect wires together that would be connected to a receptacle or fixture. Twist together all wires that will be connected to a switch. Make sure there are no exposed wires touching the boxes at any location.

If using conduit or armored cable wiring, make the first test by touching one test lead to the neutral busbar of the service entrance panel and then touching the black wire terminal of each circuit with the opposite test lead. If the bell on the tester rings, there is probably a short circuit between a white and a black wire somewhere in the circuit. Or another possibility would be a bare spot on a black wire touching the conduit or a metal box.

If the wiring is run in conduit or armored cable and the conduit or cable is to act as

Another simpler and safer method of testing employs a new device called a "receptacle circuit tester." This little device is used *after a wiring job is completed.* You simply plug it into a receptacle. By interpreting combinations of lit and unlit lights on the device, you can tell 1) if the wiring is correct, 2) if the polarity is reversed, 3) if there is an open ground, 4) if there is an open neutral, 5) if there is an open hot, 6) if the hot and ground are reversed, or 7) if the hot is on neutral and the neutral is unwired. This tester will not indicate defects such as a reversal of ground and neutral, or two hot wires in an outlet, or the quality of the ground.

If the circuit is protected by a ground-fault circuit interrupter (GFCI), you can also use a GFCI tester, after you have made tests with the receptacle tester.

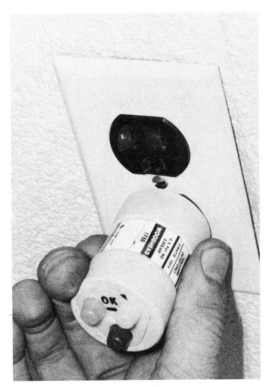

Once you have wired-in a circuit, the combinations of three lights on this tester can tell you whether the wiring is okay or whether any one of six faulty conditions exists. The label wrapped around the tester has a legend for interpretation of light combinations.

If any of the circuits are protected by a ground-fault circuit interrupter (GFCI), this GFCI tester will tell you if the GFCIs are working properly.

the ground, you should make another test to insure a proper ground. Connect the wires of the battery into white or neutral wires and then to the conduit or armored cable. Touch the leads of the bell to each white wire at every outlet and box. The bell should ring, indicating a correct ground.

If the circuit checks out in this manner, you can then disconnect all wires and connect to the proper fixtures and other hardware. Then turn on the main circuit breaker or replace the main fuse.

TROUBLESHOOTING

Start first by checking out the main fuse block or circuit breaker. Using a 2-prong fuse-and-circuit tester, perform tests as shown in photos on upcoming pages.

Locating a short circuit is even easier. A fuse that continually blows could indicate

several problems, including a defective fixture or defective wiring. To locate the problem, unplug all users such as lights and appliances. Shut off all other lights and wall
switches in the circuit involved. Screw a
100-watt light bulb into the fuse socket of
the circuit causing the problem. If the bulb
lights up and there is nothing plugged in or
turned on, the circuit has a short.

If the bulb does not light up, there is a
short in one of the items plugged or wired
into the circuit. Leave the bulb in the fuse
socket and start turning on lights, plugging
in lamps and appliances one at a time. If
the bulb in the fuse socket lights up, but the
appliance or light doesn't come on, you've
located the short.

In the fuse-type service entrance panel,

When troubleshooting an electric circuit, first look
for any blown fuses. If the fuses look okay, proceed
to the main fuse block, touching one prong of your
tester to the neutral busbar and the other to individual incoming service wires, as shown. *Take care
not to get a shock.*

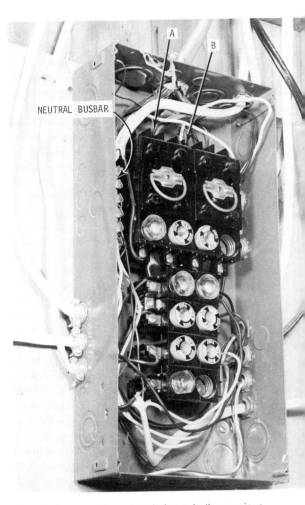

Here is the same fuse panel shown in the previous
photo. Note locations of the neutral busbar and those
of the terminals.

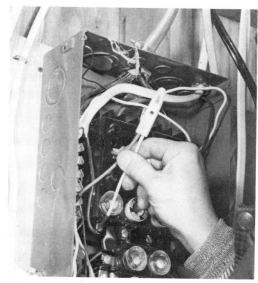

If the main and range fuse blocks are okay, check individual fuse terminals. Touch one prong of the tester to the fuse terminal and touch the other one to the neutral busbar.

If main or range fuses are bad, pull the fuse block, shown above, and replace the fuses, shown below.

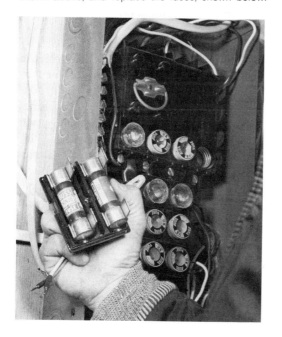

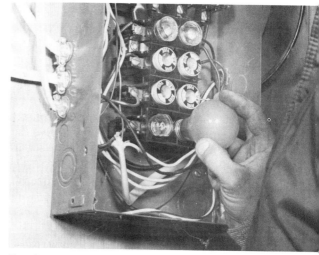

The photo above shows an easy way to locate a short circuit in a panel with fuses. First, unplug all appliances and turn off all lights on the circuit. Insert a 100-watt bulb in the fuse socket. If the bulb lights up, the circuit has a short somewhere.

you can also sometimes tell where the problem lies by the appearance of the destroyed fuse. If the fuse strip is merely melted and the window of the fuse is clear, the circuit was blown by an overload. If the window of the fuse is fogged and the tiny metal strip has burst into tiny bits, the problem is a short circuit.

Anytime you have continually blown fuses

INDICATES OVERLOADED CIRCUIT

INDICATES SHORT CIRCUIT

The face of a plug-type fuse will give you an indication of the problem in the circuit. If the window is clear and the metal strip is broken, the circuit is overloaded. If the window is fogged and the strip has burst to bits, the circuit has a short.

or tripped circuit breakers, there's trouble somewhere in your wiring, and it should be thoroughly checked out. Basically, a circuit will fail for four reasons:

1. *Loose connection in the fuse screw socket.* In this case the bottom contact on the fuse will be pitted. Remove the main fuse and tighten the screw in the bottom of the fuse socket.

2. *Improperly fitted fuse.* If a fuse doesn't reach the bottom of the socket or isn't screwed in properly, it will make poor contact. So replace it.

3. *Circuit breakers.* These rarely cause trouble. However, sometimes they become defective. And a defect in a circuit breaker can be extremely hard to track down. If all else fails in testing a circuit, the only alternative is to purchase a new circuit breaker and replace the old one.

4. *Short circuit.* As mentioned earlier, a blown fuse with a discolored window normally indicates a short circuit. To find out where the short is you can use the test mentioned earlier for a fuse panel. However, for a circuit breaker panel, turn off all lights, remove all plugs from receptacles, and trip the circuit breaker. Start turning lights back on or replacing plugs in receptacles until the circuit trips. This locates your problem.

If a circuit continues to shut off with nothing plugged in or turned on, there is a short circuit in the circuit wiring itself. Here, repair or replace the wiring.

5. *Overloaded circuit.* If the circuit stays on for some time, then shuts off again, it's overloaded. Sometimes the overload will just be temporary, caused by the starting of a motor. An electric motor draws almost three times as much current when starting as it does when running. In this case a time-delay fuse of the proper amperage will solve the problem. You can locate this type of overload quite easily by noticing which motor causes the problem. A constant overload on the other hand, means the circuit is carrying too much wattage. Perhaps you need to switch some lights or appliances to another circuit or even install a new circuit.

Index